AF411320

Friction, Vibration and Dynamic Properties of Transmission System under Wear Progression

Friction, Vibration and Dynamic Properties of Transmission System under Wear Progression

Editors

Ke Feng
Jinde Zheng
Qing Ni

MDPI • Basel • Beijing • Wuhan • Barcelona • Belgrade • Manchester • Tokyo • Cluj • Tianjin

Editors

Ke Feng
University of British
Columbia
Canada

Jinde Zheng
Anhui University of
Technology
China

Qing Ni
University of Technology
Sydney
Australia

Editorial Office
MDPI
St. Alban-Anlage 66
4052 Basel, Switzerland

This is a reprint of articles from the Special Issue published online in the open access journal *Coatings* (ISSN 2079-6412) (available at: https://www.mdpi.com/journal/coatings/special_issues/friction_vibration_dynamic).

For citation purposes, cite each article independently as indicated on the article page online and as indicated below:

LastName, A.A.; LastName, B.B.; LastName, C.C. Article Title. *Journal Name* **Year**, *Volume Number*, Page Range.

ISBN 978-3-0365-6257-5 (Hbk)
ISBN 978-3-0365-6258-2 (PDF)

Contents

About the Editors

Ke Feng

Ke Feng currently works at the University of British Columbia. He received a Ph.D. degree in mechanical engineering from the University of New South Wales, Australia, in 2021. He was a Postdoctoral Research Associate with the University of Technology Sydney in 2021. His main research interests include digital-twin-based RUL prediction, vibration analysis, structural health monitoring, dynamics, tribology, signal processing, and machine learning. He is a Fellow of Vebleo. And he is the editor and guest editor of several journals, including Mechanical Systems and Signal Processing, Frontiers in Materials, Coatings, etc.

Jinde Zheng

Jinde Zheng received the B.S. degree in mathematics from Anhui Normal University, Wuhu, China, in 2009, and the Ph.D. degree in mechanical engineering from Hunan University, Changsha, China, in 2014. From 2014 to 2016, he was a Lecturer with the Department of Mechanical Engineering, Anhui University of Technology, where he has been an Associate Professor, since 2016; he was promoted to Professor in 2021. He has authored more than 110 articles. His research interests include machinery health monitoring and fault diagnosis, statistical signal processing, and complexity theory.

Qing Ni

Qing Ni received an M.S. degree in mechanical and electrical engineering from the University of Electronic Science and Technology of China in 2019. In 2022, she submitted a Ph.D. thesis about machine condition monitoring from the University of Technology Sydney (UTS), Sydney, Australia. Her main research interests include fault diagnosis, machine learning, signal processing, and machine condition monitoring.

Editorial

Vibration-Based System Degradation Monitoring under Gear Wear Progression

Ke Feng [1], Qing Ni [2,*] and Jinde Zheng [3]

1　School of Engineering, University of British Columbia, Kelowna, BC V1V 1V7, Canada; ke.feng@ubc.ca
2　School of Mechanical and Mechatronic Engineering, University of Technology Sydney, Ultimo, NSW 2007, Australia
3　School of Mechanical Engineering, Anhui University of Technology, Maanshan 243032, China; jdzheng@ahut.edu.cn
*　Correspondence: qing.ni@outlook.com.au

Citation: Feng, K.; Ni, Q.; Zheng, J. Vibration-Based System Degradation Monitoring under Gear Wear Progression. *Coatings* **2022**, *12*, 892. https://doi.org/10.3390/coatings12070892

Received: 9 June 2022
Accepted: 21 June 2022
Published: 23 June 2022

Publisher's Note: MDPI stays neutral with regard to jurisdictional claims in published maps and institutional affiliations.

Surface wear is a common phenomenon in the service life of gear transmission systems [1]. Gear wear is the material loss from gear engaging surfaces due to sliding and rolling motions. In general, the propagation of gear wear would change the surface texture (at micro-level or macro-level), lubricating oil film thickness (if it has), and friction of the engaging gear pairs. The change in surface texture and friction can directly impact the gear contact load distribution and stress concentration. Consequently, the gear wear propagation can accelerate the occurrence of some functional failures and lead to the unexpected shutdown of the gear transmission system, which could cause enormous economic loss. Therefore, it is vital to monitor the gear wear progression so that reliable predictive maintenance-based decisions can be made to ensure a safe operation of the gear transmission system.

In general, the vibration signal measured from the gear transmission system can be affected by the surface wear progression [2]. The reason is that the gear mesh excitation and gear transmission ratio are sensitive to the gear surface geometry [3]. Therefore, vibration analysis can be a powerful tool for monitoring gear wear propagation. Also, compared with the wear particle analysis, vibration analysis has its unique advantages [4]. For example, the vibration characteristics can indicate the instant gear transmission performance change caused by gear wear. Thus, it can avoid the delay caused by the process progression of wear particle analysis. Therefore, vibration-based gear wear monitoring has attracted significantly increasing attention from researchers in recent decades. This editorial will give an overview of the development of vibration-based gear wear monitoring.

Some research works use the vibration features to monitor the system degradation behaviors caused by gear wear. For instance, as a prevalent health monitoring indicator, the gear meshing harmonic was applied to track the gear wear propagation in reference [5]. Also, the research [5] revealed that the higher-order gear meshing harmonics are closely relevant to the gear wear propagation progression. A similar conclusion was also drawn in [6]. In addition, the features of the cepstrum of vibrations were utilized for gear wear monitoring in [6], and the experimental results prove the effectiveness of the features from the spectrum and cepstrum in gear wear monitoring. However, the system degradation behaviors caused by gear wear propagation are highly complex. The dynamic interaction of multiple gear wear mechanisms and dynamic characteristics makes the wear propagation rate and patterns not constant. As a result, limited gear meshing harmonics do not have the capability of fully revealing the gear wear propagation status. Therefore, multiple gear meshing harmonics were included in [7] to track the gear wear propagation progression. In [7], the gear meshing harmonics were selected by a specific rule to derive two new gear wear monitoring indicators. With the help of the two developed gear wear monitoring indicators, the accumulated degradation behaviors and instant health status change induced by gear wear can be well indicated. Two endurance tests were arranged to demonstrate and validate the performance of the two gear wear monitoring indicators [7]. In addition to the gear meshing harmonic, some other techniques were also employed for gear wear monitoring. For example, the bispectrum of vibration signal was used in [8] to evaluate the gear wear progression. Based on the modulation signal bispectrum, the authors proposed a sideband estimator in [8], and its effectiveness was verified using the signals from a two-stage gearbox system.

In general, gear meshing harmonics and the bispectrum are closely relevant to the wear-induced gear tooth profile change. In the progression of gear wear, the micro-level surface morphology also changes significantly. To quantify the gear wear-induced micro surface morphology change, the cyclostaionary analysis was applied and proved to be a promising tool. The relationship between surface roughness and the second-order cyclostaionary (CS2) indicator was briefly investigated in [9,10]. However, due to the complex gear wear propagation process and the limitation of the experimental arrangement, insufficient conclusions were drawn in [9,10]. Later, the internal relationship of gear wear tribological features with the measured vibrations were proposed and proved in the study [11], as shown in Equation (1)

$$f_v \propto v_s \cdot f_s \tag{1}$$

where f_v (Hz) represents the frequency of the dominant sliding induced vibration signal, v_s (m/s) denotes the sliding velocity of the engaging gear surface, and the dominant spatial frequency of the engaging surface morphology is written as f_s (1/m). In the research [11], the wear mechanism identification was first achieved using the online approach. More specifically, the abrasive wear and fatigue pitting were identified using the cyclostationary properties of the measured vibration signal, and the corresponding wear evolutions were accurately tracked. Inspired by the research [11], the capability of another cyclostationary analysis tool (cyclic-correntropy) in gear wear monitoring was evaluated in [12], and a novel gear wear monitoring indicator, namely the weighted cyclic correntropy operator, was proposed. Some endurance tests were conducted to demonstrate the performance of the developed gear wear monitoring indicator.

The above-discussed research works have contributed to the development of vibration-based gear wear monitoring and wear mechanism identification, significantly benefiting nondestructive gear wear monitoring and ensuring the safe operation of the gear transmission system. However, the characteristics of the vibration signal can only reflect the degradation trend of the gear transmission system, and the details of the contact status of the engaging gear can not be revealed. The knowledge of the gear contacting status is valuable for gear maintenance and gear design/optimization. The digital twin is an emerging technique for gear health management and has attracted significant attention from the research community and industry practices [13–15]. The digital twin model can help reflect the degradation status of the gear transmission system during wear progression, which can help the analyst understand the in-depth degradation mechanisms. Also, the remaining useful life (RUL) of the gear transmission system can be well predicted using the digital twin models. Therefore, the digital twin technique is of high practical value to gear wear monitoring and gear health management.

Based on the frame of the digital twin technique, the abrasive wear-induced tooth profile change was well predicted in [16]. In research [16], the root mean square (RMS) of the measured vibration signal was compared with the simulation signal to help update and calibrate the wear coefficient of the Archard wear model. Through the real-time communication between measurements of the physical structure and simulation models, the wear propagation rate change can be timely captured so that the abrasive wear-induced system degradation behaviors can be well indicated and the RUL of the gear system can be accurately predicted. Later, an improved Archard wear model was developed in [17], and this improved Archard wear model has been integrated into the digital twin frame for gear RUL prediction. Two run-to-failure tests with different lubrication conditions were used to verify the effectiveness of the developed vibration-based updating (digital twin) scheme for abrasive wear propagation prediction. The above two methodologies focus on the abrasive wear progression. However, in practice, multiple wear mechanisms interact with each other, resulting in highly complex system degradation progress [18,19]. Therefore, the technique which can handle multiple wear mechanisms is of great value for industrial practices. To this end, a novel fatigue pitting model was developed in [20] based on the Lundberg-Palmgren fatigue theory [21]. Then the dynamic model, developed fatigue pitting model, and Archard wear model were combined into twin models of the gear transmission system in [20]. To realize the real-time connection of twin models and the physical structure of the gear transmission system, a novel updating scheme, including RMS and second-order cyclostaionary indicator ICS2 [22], was also proposed to update the parameters of the twin models if necessary. The developed digital twin-based gear wear propagation prediction performance was illustrated and proved by two endurance tests [20].

Even though the above introduced digital twin-based gear wear methodologies have achieved satisfying prediction results, the prediction accuracy is highly enslaved to the fidelity of the digital twin models. Establishing a high-fidelity digital twin model is very costly, and high-level expert knowledge is required. Therefore, the intelligent digital twin model establishment technique/methodology is in vital need, which can help increase the practical value of the digital twin techniques for gear wear monitoring. Moreover, the digital twin technique will be a research hotspot in the area of wear analysis and gear health management.

Conflicts of Interest: The authors declare no conflict of interest.

References

1. Ding, H.; Kahraman, A. Interactions between nonlinear spur gear dynamics and surface wear. *J. Sound Vib.* **2007**, *307*, 662–679. [CrossRef]
2. Randall, R.B. *Vibration-Based Condition Monitoring: Industrial, Automotive and Aerospace Applications*; John Wiley & Sons: Hoboken, NJ, USA, 2021.
3. Wojnarowski, J.; Onishchenko, V. Tooth wear effects on spur gear dynamics. *Mech. Mach. Theory* **2003**, *38*, 161–178. [CrossRef]
4. Ni, Q.; Ji, J.; Feng, K. Data-driven prognostic scheme for bearings based on a novel health indicator and gated recurrent unit network. *IEEE Trans. Ind. Inform.* **2022**, 1. [CrossRef]
5. Randall, R. A New Method of Modeling Gear Faults. *J. Mech. Des.* **1982**, *104*, 259–267. [CrossRef]
6. Žiaran, S.; Darula, R. Determination of the State of Wear of High Contact Ratio Gear Sets by Means of Spectrum and Cepstrum Analysis. *J. Vib. Acoust.* **2013**, *135*, 021008. [CrossRef]
7. Hu, C.; Smith, W.A.; Randall, R.B.; Peng, Z. Development of a gear vibration indicator and its application in gear wear monitoring. *Mech. Syst. Signal Process.* **2016**, *76–77*, 319–336. [CrossRef]
8. Zhang, R.; Gu, X.; Gu, F.; Wang, T.; Ball, A.D. Gear Wear Process Monitoring Using a Sideband Estimator Based on Modulation Signal Bispectrum. *Appl. Sci.* **2017**, *7*, 274. [CrossRef]
9. Yang, Y.; Smith, W.A.; Borghesani, P.; Peng, Z.; Randall, R.B. Detecting changes in gear surface roughness using vibration signals. In Proceedings of the Acoustics Conference, Hunter Valley, NSW, Australia, 15–18 November 2015; pp. 1–10.
10. Zhang, X.; Smith, W.A.; Borghesani, P.; Peng, Z.; Randall, R.B. *Use of Cyclostationarity to Detect Changes in Gear Surface Roughness Using Vibration Measurements, Asset Intelligence through Integration and Interoperability and Contemporary Vibration Engineering Technologies*; Springer: Cham, Switzerland, 2019; pp. 763–771.
11. Feng, K.; Smith, W.A.; Borghesani, P.; Randall, R.B.; Peng, Z. Use of cyclostationary properties of vibration signals to identify gear wear mechanisms and track wear evolution. *Mech. Syst. Signal Process.* **2021**, *150*, 107258. [CrossRef]
12. Feng, K.; Ji, J.; Li, Y.; Ni, Q.; Wu, H.; Zheng, J. A novel cyclic-correntropy based indicator for gear wear monitoring. *Tribol. Int.* **2022**, *171*, 107528. [CrossRef]
13. Tao, F.; Zhang, H.; Liu, A.; Nee, A.Y.C. Digital Twin in Industry: State-of-the-Art. *IEEE Trans. Ind. Inform.* **2019**, *15*, 2405–2415. [CrossRef]
14. Liu, M.; Fang, S.; Dong, H.; Xu, C. Review of digital twin about concepts, technologies, and industrial applications. *J. Manuf. Syst.* **2021**, *58*, 346–361. [CrossRef]
15. Errandonea, I.; Beltrán, S.; Arrizabalaga, S. Digital Twin for maintenance: A literature review. *Comput. Ind.* **2020**, *123*, 103316. [CrossRef]
16. Feng, K.; Borghesani, P.; Smith, W.A.; Randall, R.B.; Chin, Z.Y.; Ren, J.; Peng, Z. Vibration-based updating of wear prediction for spur gears. *Wear* **2019**, *426–427*, 1410–1415. [CrossRef]

17. Feng, K.; Smith, W.A.; Peng, Z. Use of an improved vibration-based updating methodology for gear wear prediction. *Eng. Fail. Anal.* **2021**, *120*, 105066. [CrossRef]
18. Weibring, M.; Gondecki, L.; Tenberge, P. Simulation of fatigue failure on tooth flanks in consideration of pitting initiation and growth. *Tribol. Int.* **2019**, *131*, 299–307. [CrossRef]
19. Morales-Espejel, G.; Rycerz, P.; Kadiric, A. Prediction of micropitting damage in gear teeth contacts considering the concurrent effects of surface fatigue and mild wear. *Wear* **2018**, *398–399*, 99–115. [CrossRef]
20. Feng, K.; Smith, W.A.; Randall, R.B.; Wu, H.; Peng, Z. Vibration-based monitoring and prediction of surface profile change and pitting density in a spur gear wear process. *Mech. Syst. Signal Process.* **2022**, *165*, 108319. [CrossRef]
21. Lundberg, G. Dynamic capacity of rolling bearings. *IVA Handl.* **1947**, *196*, 12. [CrossRef]
22. Raad, A.; Antoni, J.; Sidahmed, M. Indicators of cyclostationarity: Theory and application to gear fault monitoring. *Mech. Syst. Signal Process.* **2008**, *22*, 574–587. [CrossRef]

Article

A Dynamic Wear Prediction Model for Studying the Interactions between Surface Wear and Dynamic Response of Spur Gears

Jinzhao Ren [1,*] and Huiqun Yuan [2]

[1]   School of Mechanical Engineering and Automation, Northeastern University, Shenyang 110819, China
[2]   School of Science, Northeastern University, Shenyang 110819, China
*   Correspondence: rjc626@hotmail.com

Abstract: Surface wear, as a major failure mode of gear systems, is an unavoidable phenomenon during the whole life of gears. It also induces other gear damages, such as fatigue cracks, surface pitting and spalling. Ultimately, those defects may result in the sudden failure of a gearbox transmission system, which can lead to a serious accident and unexpected economic loss. Therefore, it can provide huge cost and safety benefits to industries to monitor gear wear and predict its propagation. Gear wear raises the error rate of gear transmission systems, typically leading to improvements in dynamic loads, vibration, and noise. In return, the increased load conversely aggravates wear, creating a feedback cycle between dynamic responses and surface wear. For this purpose, a wear prediction model was incorporated into a tribo-dynamic model for quantitatively investigating how surface wear and gear vibration are mutually affected by each other. To obtain more precise dynamic responses, the tribo-dynamic model integrates the time-varying mesh stiffness, load-sharing ratio and friction parameters. To improve the computational efficiency and guarantee the calculation precision, an improved and updated wear depth methodology is constructed in the wear prediction model. This paper demonstrates the capability of the proposed dynamic wear prediction model in the investigation of the interaction effects between gear dynamics and surface wear, allowing for the development of improved gear wear prediction tools. The obtained results indicate that the surface wear impacts the dynamic characteristics, even with slight wear. In the initial stage of wear, the friction coefficient decreases slightly, largely due to the reduction in surface roughness; but the friction force increases because of the improved dynamic meshing force. Although the initial wear depth distributions of a pinion under dynamic and static conditions are similar, the wear depth distributions under dynamic conditions becomes significantly different compared to the those under static conditions with the wear process. The maximum wear depth of a pinion under dynamic conditions is about 1.6 times as the corresponding static conditions, when the wear cycle comes to 4×10^4. Similarly, the maximum accumulative wear depth of a pinion under dynamic conditions reaches 1.2 times of that under static conditions. Therefore, the proposed dynamic wear prediction model is more appropriate to be applied to the surface wear of gears.

Keywords: coupling effects; dynamic model; gear wear; wear prediction

Citation: Ren, J.; Yuan, H. A Dynamic Wear Prediction Model for Studying the Interactions between Surface Wear and Dynamic Response of Spur Gears. *Coatings* **2022**, *12*, 1250. https://doi.org/10.3390/coatings12091250

Academic Editor: Giuseppe Carbone

Received: 4 August 2022
Accepted: 24 August 2022
Published: 26 August 2022

1. Introduction

Gear transmission is an important form of mechanical transmission, and the reliability and durability of gears are critical to the total life of mechanical equipment. Due to its special mechanical structure, a gear system is used in wide range of mechanical systems, including the mining industry, helicopters, and wind turbines. In practice, a gearbox often operates under harsh working conditions. Consequently, the inevitable gearbox failures frequently result in serious accidents and unforeseen financial losses. Gear wear is an unavoidable phenomenon in the service life of gear. It will cause the development of stress concentrations, and serve as initiation sites for other modes of gear failure, for example, scuffing, macro-pitting, and gear cracking [1], which could cause the vibration characteristics to change significantly. Thus, the monitoring and predicting of gear surface wear is vital for the health management of the gear system.

In theory, surface wear and dynamic characteristics of gear affect each other. The surface wear caused by relative sliding modifies the geometry of gear tooth profile. Therefore, the gear transmission error would vary dynamically [2], especially for spur gears, whose gear transmission error is susceptible to surface wear [3]. At the same time, the dynamic characteristics of the gearbox will be altered accordingly. Consequently, vibration responses are expected to rise in level. Surface wear is closely related to contact pressure which depend on dynamic meshing force [4]. Thus, the wear process also is promoted because of improved dynamic meshing force. This two-way relationship between them will yield more complex gear dynamic characteristics and makes monitoring the condition of gear wear more difficult than other failures.

However, existing researches mainly focus on investigating the impact of gear wear on dynamic responses [3,5–8]. In comparison, investigations on the effects of dynamic behaviors on gear wear are quite few [9–11]. Some studies which predict gear wear by a quasi-static wear model have been published [12–14]. Nevertheless, meshing force and sliding velocity under dynamic conditions are significantly different from the ones under quasi-static conditions. Therefore, a comprehensive dynamic wear prediction model for analyzing their coupling effects is vitally needed. In this paper, the contact pressure and sliding velocity from the dynamic model are fed into wear prediction model to determine the wear depth of each engagement point. Subsequently, the gear tooth profile is renewed in the dynamic model by feeding an updated geometric transmission error, which characterizes the deviation of the profile from the ideal involute curve. Then new vibration responses are acquired from the dynamic model again. This cycle repeats to produce estimated gear wear profiles, as well as corresponding simulated vibration responses, showing clearly how gear wear and dynamic responses affect each other.

Understanding the effort of gear wear on dynamic responses is essential to track gear wear evolution, and the dynamic vibration analysis is a widely used application and effective technique to monitor and predict gear surface wear [15,16]. The dynamic characteristics of a gear system are theoretically susceptible to deviations of the tooth surface from a perfect involute [17]. Surface deviation mainly includes wear-induced geometric deviation and elastic deviation caused by meshing stiffness and contact force. According to this theory, Ding and Kahraman [9] described the impact of dynamic response on surface wear in terms of an external displacement excitation and a periodically time-varying meshing stiffness (TVMS) function. They utilized a single degree of freedom (DOF) torsional model and incorporated it with a wear prediction model [13] to research the interactions between the wear process and dynamic characteristics of gear system. The simulation analysis demonstrated the interaction behaviors between gear wear and dynamic characteristics. The accurate prediction for dynamic behaviors of gear system requires a thorough dynamic model. However, the dynamic models employed in [9,10] solely contained the torsional degrees of freedom. As a result, the prediction accuracy of the dynamic analysis is reduced because of it ignoring the translational deflections of the shaft bending and bearing radial. In order to address this issue, ref. [11] applied a 3 DOFs model including both torsional and translational motions of gears to acquire dynamic responses of a gear system for further wear analysis. Nevertheless, empirical formulas were applied to calculate the meshing stiffness and transmission error, which also could degenerate the accuracy of obtained responses. In this paper, the potential energy method was adopted to estimate the meshing stiffness. This method has been widely implemented to analytically model the TVMS of gear pairs because of its precision. Moreover, it is convenient to consider the impact of surface wear on meshing stiffness by this method when severe wear occurs.

In addition, wear alters the gear surface quality, such as hardness and roughness, which can significantly impact the tribological behavior of the meshing gear pair. J. Jamari et al. [18] measured the surface roughness values Ra of gear before and after running-in tests. They found that the slight change in roughness magnitude would contribute much to gear wear during the running-in process. The friction behavior at the macroscopic length scale is highly correlated with the microscopic topography of the contacting surfaces. Thus, the wear-induced reduction in roughness should be considered, and friction, which is recognized as an essential source of gear vibrations, should be incorporated into the dynamic model to account for the effects of gear wear. Although much research [19–23] incorporated the effects of frictional force into gear dynamic model to investigate the impact of friction on gear vibration, the impact of surface wear on friction is disregarded. In this study, a friction model is incorporated in the dynamic model to investigate the impact of wear on friction and further on dynamic responses.

Our review of the aforementioned studies reveals that the existing dynamic models are insufficient for studying the interactions between gear tooth surface wear and gear dynamic characteristics due to the lack of a comprehensive dynamic wear prediction model. Therefore, this paper aims at establishing a thorough dynamic wear prediction model by incorporating a wear model into a tribo-dynamic model. To guarantee a reliable and accurate dynamic wear model, an 8 DOFs model, which integrates TVMS, dynamic friction parameters, and time-varying load-sharing ratio, is established. Then, it is combined with Archard wear model. An improved wear depth updated methodology is constructed in the wear prediction model. With the proposed dynamic wear prediction model, the interaction effects between gear tooth surface wear and gear dynamic characteristics can be studied.

In the following, Section 2 firstly introduces the proposed 8 DOFs translation-rotational-coupled nonlinear dynamic model. An improved dynamic wear prediction scheme is proposed in Section 3. In the Section 4, the simulations are executed and results are discussed. Conclusion and further study are made in Section 5.

2. Dynamic Model Development

The dynamic model is to obtain the dynamic responses to be input to the wear model. In this section, an 8 DOFs tribo-dynamic model of spur gear system is developed by incorporating TVMS, time-varying load sharing, dynamic friction coefficient; see Figure 1.

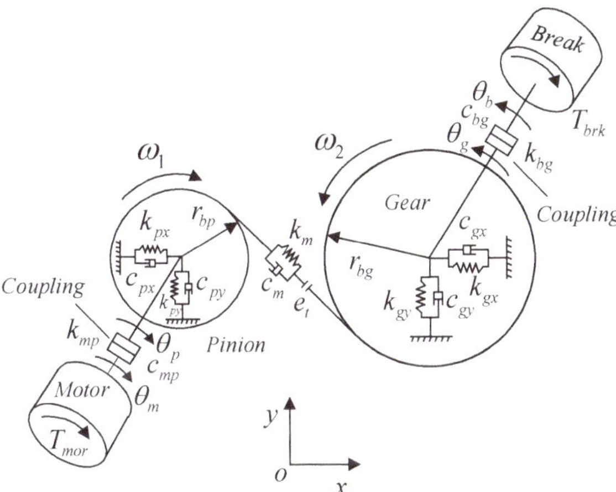

Figure 1. Diagram of the gear dynamic model.

The equations of motion used to describe the coupled torsional and translational models can be formulated as:

$$\begin{cases} m_p\ddot{y}_p + c_{py}\dot{y}_p + k_{py}y_p + F_{pg}\cos\varphi + f_p\sin\varphi = 0 \\ m_g\ddot{y}_g + c_{gy}\dot{y}_g + k_{gy}y_g - F_{pg}\cos\varphi + f_g\sin\varphi = 0 \\ m_p\ddot{x}_p + c_{px}\dot{x}_p + k_{px}x_p + F\sin\varphi + f_p\cos\varphi = 0 \\ m_g\ddot{x}_g + c_{gx}\dot{x}_g + k_{gx}x_g - F\sin\varphi + f_g\cos\varphi = 0 \\ J_m\ddot{\theta}_m + c_{mp}(\dot{\theta}_m - \dot{\theta}_p) + k_{mp}(\theta_m - \theta_p) - T_{mor} = 0 \\ J_p\ddot{\theta}_p - c_{mp}(\dot{\theta}_m - \dot{\theta}_p) - k_{mp}(\theta_m - \theta_p) + r_{pb}F + M_{pf} = 0 \\ J_g\ddot{\theta}_g - c_{bg}(\dot{\theta}_b - \dot{\theta}_g) - k_{bg}(\theta_b - \theta_g) - r_{gb}F + M_{gf} = 0 \\ J_b\ddot{\theta}_b + c_{bg}(\dot{\theta}_b - \dot{\theta}_g) + k_{bg}(\theta_b - \theta_g) + T_{brk} = 0 \end{cases} \tag{1}$$

Herein, θ_p, θ_g, θ_b, and θ_m refer to the rotational displacements of the pinion, gear, load, and motor, respectively; and x_p, y_p and x_g, y_g indicate the vibration displacements of the pinion and gear, respectively. m_p and m_g donate the masses of the pinion and gear correspondingly. Likewise, the inertias of the pinion and the gear are expressed as J_p and J_g, respectively.

The major parameters contained in the model are given in Table 1. The contact force $F(t)$ is modelled by gear meshing stiffness k_m, meshing damping c_m and geometric transmission error (GTE)e, see Equation (2).

$$\begin{aligned} F(t) \quad &= k_m(t)\left[r_{pb}\theta_p - r_{gb}\theta_g + (x_p - x_g)\sin\varphi + (y_p - y_g)\cos\varphi) + e_t\right] \\ &+ c_m(t)\left[r_{pb}\dot{\theta}_p - r_{gb}\dot{\theta}_g + (\dot{x}_p - \dot{x}_g)\sin\varphi + \left(\dot{y}_p - \dot{y}_g\right)\cos\varphi + \dot{e}_t\right] \end{aligned} \tag{2}$$

where r_{pb} and r_{gb} represents the base circle radius of the pinion and gear, φ indicates the pressure angle.

Table 1. Parameters of the gear system.

Parameters	Pinion	Gear
Gear type	Standard involute, full teeth	
Material	S45 C	
Modulus of elasticity, E	205 Gpa	
Poisson's ratio, v	0.3	
Face width, W	20 mm	
Module, M	2 mm	
Pressure angle	20°	
Addendum	1.00 mm	
Dedendum	1.25 mm	
Number of teeth Z_1, Z_2	19	31
Pitch radius, r/mm	19	31

The total frictional torques, namely M_{pf} and M_{gf}, can be given as

$$\begin{cases} M_{pf} = \sum\limits_{i=1}^{n} r_{pi}f_{pi} \\ M_{gf} = \sum\limits_{i=1}^{n} r_{gi}f_{gi} \end{cases}, \quad n = 1, 2 \tag{3}$$

where r_{pi}, r_{gi} (i = 1, 2) denote the radii of curvature at the contact points, and can be expressed as

$$\begin{cases} r_{pi} = r_{pb}\left(\theta_{ini} + (i-1)\frac{2\pi}{z_1} + \theta_p\right), \\ r_{gi} = L - r_{pi} \end{cases} \quad i = 1, 2 \tag{4}$$

where i represents the ith gear tooth; L denotes the length of the line of action (LOA), computed as $L = \left(r_{pb} + r_{gb}\right)\tan\varphi$. Therefore, the instantaneous tooth surface velocities u_{pi} and u_{pi} can be expressed as

$$\begin{cases} u_{pi} = \omega_p \cdot r_{pi} \\ u_{gi} = \omega_g \cdot r_{gi} \end{cases}, \quad i = 1,2 \tag{5}$$

Herein, ω_p and ω_g represent the rotational speeds of the pinion and gear, respectively. The relative velocity $u_s = u_{pi} - u_{gi}$ will be used to calculate the sliding distance for wear model and determine the direction of friction force. When $u_s > 0$, f_p is negative; conversely, when $u_s < 0$, f_p is positive with $u_s = 0$, $f_p = 0$. The friction force between the meshing teeth can be calculated by:

$$f_{pi} = -f_{gi} = \mu_i \kappa_i F_{pg}, i = 1,2 \tag{6}$$

where κ_i and μ_i denote the load sharing ratio (LSR) and dynamic friction coefficient, respectively. To compute the load distribution between the meshing teeth pairs, we used a load distribution model proposed by Pedrero [24]. The detailed descriptions can be referred to reference [24].

$$\kappa_i = \begin{cases} \frac{1}{3}\left(1 + \frac{\xi - \xi_{ini}}{\xi_a - \xi_{ini} - 1}\right) & \xi_{ini} \le \xi \le \xi_a - 1 \\ 1 & \xi_a - 1 \le \xi \le \xi_{ini} + 1 \\ \frac{1}{3}\left(1 + \frac{\xi - \xi_a}{\xi_{ini} + 1 - \xi_a}\right) & \xi_{ini} + 1 \le \xi \le \xi_a \end{cases}, \tag{7}$$

where $\xi_{ini} = \frac{z_1 \theta_{ini}}{2\pi}$, $\xi_a = \xi_{ini} + \varepsilon$, and ε denotes the contact ratio.

The friction model proposed by Xu et al. [25] is applied to obtain the dynamic friction coefficient. This model is on the basis of regression of experimental tests under a wide range of operating conditions and has been demonstrated by both simulated and experimental data. The dynamic friction coefficient is given as

$$\mu_i = e^f P_h^{b_2} |S_r|^{b_3} V_e^{b_6} v_0^{b_7} R^{b_8}, \tag{8}$$

where

$$\begin{cases} f = b_1 + b_4 |S_r| P_h log_{10}(v_0) + b_5 e^{-|S_r| P_h log_{10}(v_0)} + b_9 e^s \\ S_r = \frac{2(u_{pi} - u_{gi})}{(u_{pi} + u_{gi})}, \\ V_e = 0.5(u_{pi} + u_{gi}) \end{cases} \tag{9}$$

and where P_h is Hertzian contact pressure, and S_r, V_e, v_0 donates slide to roll ratio, entrainment velocity, and the inlet oil viscosity, respectively. S is the RMS composite surface roughness, and b_1 to b_9 are regression coefficients, with $b_{1-9} = -8.92, 1.03, 1.04, -0.35, 2.81, -0.10, 0.75, -0.39, 0.62$.

In this study, the potential energy method, which includes the bending, shear, axial compressive, and Hertzian contact energies along with the fillet foundation deflection was adopted to estimate the meshing stiffness [26]. According to the potential energy method, the bending, shear, and axial compressive potential energies stored in the meshing teeth can be calculated by [27]:

$$\begin{cases} U_b = \frac{F^2}{2k_b} \\ U_s = \frac{F^2}{2k_s} \\ U_a = \frac{F^2}{2k_a} \end{cases} \tag{10}$$

where F denotes the gear contact force, and k_b, k_s, and k_a are the bending, shear, and axial compressive stiffnesses, respectively. Based on the results derived by Yang and Sun [28], the Hertzian contact stiffness for gear pairs can be linearized to a constant and determined using the tooth width Wand material properties, namely the Poisson's ratio (ν) and elastic modulus (E).

$$k_h = \frac{\pi EW}{4(1 - \nu^2)} \tag{11}$$

With Hertzian, bending, shear and axial compressive stiffness, the gear mesh stiffness for one tooth pair can be obtain by using [25]:

$$\frac{1}{k_m} = \frac{1}{k_h} + \sum_{j=1}^{2}\left[\frac{1}{k_{b,j}} + \frac{1}{k_{s,j}} + \frac{1}{k_{a,j}}\right] \tag{12}$$

where $j = 1, 2$ donates the pinion and gear, respectively. For two pairs of meshing gears, the total effective mesh stiffness can be calculated as

$$\frac{1}{k_m} = \sum_{j=1}^{2}\left[\frac{1}{k_{h,j}} + \frac{1}{k_{b1,j}} + \frac{1}{k_{s1,j}} + \frac{1}{k_{a1,j}} + \frac{1}{k_{b2,j}} + \frac{1}{k_{s2,j}} + \frac{1}{k_{a2,j}}\right], \tag{13}$$

where $j = 1$ and 2 denote the first and second meshing teeth pair, respectively. The mesh stiffness with angular displacement is demonstrated in Figure 2.

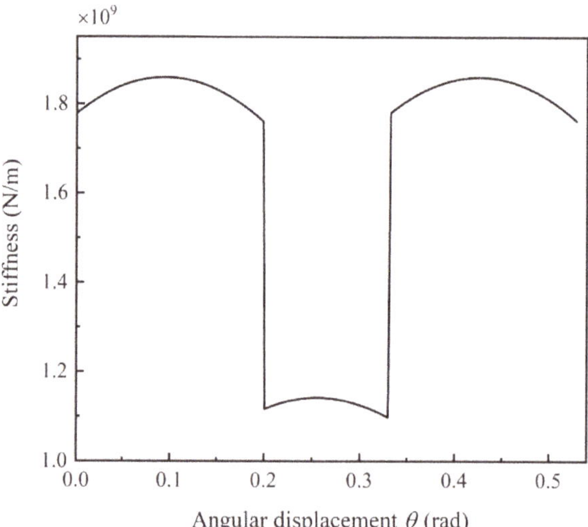

Figure 2. Time-varying mesh stiffness (TVMS) with angular displacement.

On account of the gear wear induced tooth profile change h is in micron level, the wear induced meshing stiffness change is around in the 10^{-18} N/m level [29], while the meshing stiffness is around in the 10^{9} N/m level correspondingly. Therefore, the stiffness change can be neglected compared with the value of meshing stiffness.

3. Dynamic Wear Prediction Model

In this section, an improved dynamic wear prediction model for spur gear system is proposed. The well-known Archard wear equation [30], which is one of the earliest wear laws, is used in this work. The Archard wear model is the most commonly applied in gear wear, because it takes into account the contact pressure, sliding condition and material properties of gear contact surface. The Archard's wear equation is generally expressed as:

$$dH^i = kP_h^i ds^i \tag{14}$$

where H stands for the wear depth, k is a dimensionless wear coefficient, s is the sliding distance, and P_h is the hertzian contact pressure between the mating point on the mating gear. The contact pressure P_h can be calculated as:

$$P_h^i = \sqrt{\frac{F_t^i \kappa_i E}{2b\pi r_{eq}^i}},$$ (15)

where b denotes the gear width, E denotes elastic modulus, and r_{eq} represents the equivalent radius of curvature. Consequently, the predicted wear depth accumulated on the pinion and gear can be expressed:

$$\begin{cases} H_{p,n+1}^i = H_{p,n}^i + kP_h^i ds^i \\ H_{g,n+1}^i = \frac{Z_1}{Z_2} H_{p,n}^i + kP_h^i ds^{i\,'} \end{cases}$$ (16)

Herein, $H_{p,n+1}$, $H_{g,n+1}$ represent the accumulated wear depth after the certain cycle n of each meshing points on the pinion and gear. However, it is almost impossible that the values of wear depths $H_{p,n+1}$, $H_{g,n+1}$ are updated after each loading cycle with the evolution of gear surface conditions. Therefore, a modified method shown below is used to improve the computational efficiency and guarantee the calculation precision.

$$\begin{cases} H_{p,N+1}^i = H_{p,N}^i + mkP_h^i ds^i \\ H_{g,N+1}^i = H_{g,N}^i + \frac{Z_1}{Z_2} mkP_h^i ds^{i\,'} \end{cases}$$ (17)

According to Equation (14), the wear depth ΔH^i after one wear cycle can be determined using sliding distance s, dynamic contact pressure P_h and wear coefficient k [12,31]. The wear depth ΔH^i of each contact point are assumed to remind the same during experiencing a fixed wear cycle m. When the maximum accumulated wear depth of any point on the mating surface during the fixed wear cycle m meets a predetermined wear threshold ε^m, the gear surface needs to be renewed in order to update contact pressure by performing another dynamic analysis of gear system. The total cumulative wear depth of every point on the surface is obtained by summing up wear depths of every point for all processes with different pressure updates. Figure 3 demonstrates the flowchart of the dynamic wear prediction model.

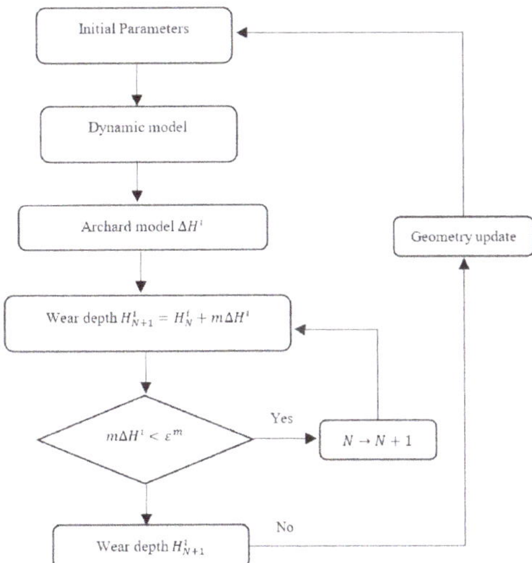

Figure 3. Whole procedure of the dynamic wear prediction model.

4. Numerical Results and Discussion

In this section, with the dynamic gear model and wear prediction method introduced in Section 3, the coupling effects between the gear surface wear and the gear dynamic characteristics will be investigated. GTE is the geometric deviation from perfect gear, and it can be used to represent gear surface wear in subsequent analysis. With valid evaluated meshing stiffness and GTE, the gear mesh force can be calculated by Equation (2), then the wear induced gear dynamic responses and vibrations can be achieved through the proposed dynamic model. Table 2 lists the basic parameters of the spur gear pair and dynamic simulation.

Table 2. Parameters of the spur gear transmission involved in the dynamic simulation.

Parameters	Values
Mass of the pinion, m_p	0.7 kg
Mass of the gear, m_g	1.822 kg
Mass moment inertia of the pinion, J_p	2.331×10^{-4} kg·m^2
Mass moment inertia of the gear, J_g	1.392×10^{-3} kg·m^2
Mass moment inertia of the motor, J_m	2.1×10^{-3} kg·m^2
Mass moment inertia of break, J_b	1.05×10^{-2} kg·m^2
Damping coefficient of bearing, $c_{bx}, c_{by}, c_{px}, c_{py}$	4×10^5 N·s/m
Stiffness of bearing, $k_{bx}, k_{by}, k_{px}, k_{py}$	5×10^8 N·m
Damping of coupling, c_{mp}, c_{bg}	3×10^4 N·s/m
Coupling stiffness, k_{mp}, k_{bg}	4×10^7 N·m

4.1. Effects of Gear Surface Wear on the Dynamic Response

In this case, the rotational speed of input shaft is set as 30 Hz and the torque of brake is set as 60 Nm. For comparing convenience, gear maximum wear depth with 0.1 μm, 0.5 μm, and 1 μm are considered in this study. Gear surface wear was found to affect the dynamic signals of the gearbox, which resulting in higher transmission error and then higher vibration and noise, as shown in Figure 4. With respect to the dynamic transmission error, the maximum values are 16.70 μm, 18.08 μm and 19.47 μm. Figure 5 depicts the vertical velocity $\dot{y}_p$ of a pinion at difference wear levels. Correspondingly, the root-mean-square (RMS) value of $\dot{y}_p$ is 1.1mm/s with 0.1 μm wear, then increases to 1.19mm/s with 0.5 μm wear and reaches to 1.28mm/s with 1 μm wear. Therefore, gear surface wear serves as one kind of geometric deviation, which can increase transmission error of the gear system resulting in higher vibration. In addition, Figure 6 illustrates frequency spectrum of $\dot{y}_p$ with 0.5 μm wear, and the change trends of first four meshing frequency (MF) amplitudes under different wear severities are plotted in Figure 7. It is observed that the amplitudes of MF will increase alone with gear wear process. Therefore, spectrum can be used for monitoring gear wear process.

The predicted gear meshing force F under different wear severities are displayed in Figure 8. As demonstrated in Figure 8, the meshing fore with 1 μm wear has higher fluctuation than those with 0.1 μm and 0.5 μm wear. The mean values of mesh force are calculated, which are 3311.2 N, 3431.9 N and 3697.6 N under different wear severities. Therefore, gear wear increases the magnitude of gear meshing force. According to Equation (2), the increase in GTE caused by tooth surface wear inevitably enhance the dynamic meshing force. Apart from that, friction is a crucial source of gear vibrations. The change in friction during the wear process also affects the meshing force. Figures 9 and 10 demonstrate the evolution of coefficient of friction (COF) and friction force with wear process. It should be pointed out that, for clarity, Figures 9 and 10 only tracked the COF and friction force between the first mating pair which initially mesh at gear tooth root. As shown in Figure 9, the COF decreases slightly with running-in wear, mainly because of the reduction in surface roughness value. J. Jamari et al. [18] found that this reduction in roughness still exist even after running-in wear, and becomes higher as running-in time. Therefore, in this simulation, the initial RMS composite surface roughness is set as 2.1 μm, and it changes to 1.9 μm with

0.5 µm wear and 1.5 µm with 1 µm wear. The results depicted in Figure 9 reveal that the COF is sensitive to the change in roughness magnitude during the running-in wear process. However, the friction force increases as shown in Figure 10 because of the improved dynamic meshing force, especially during the single-tooth-pair meshing duration. Based on Equation (1), the increasing friction force leads to the increase in dynamic responses and meshing force, and it is believed that the effect of friction will become more and more significant as the wear depth increases.

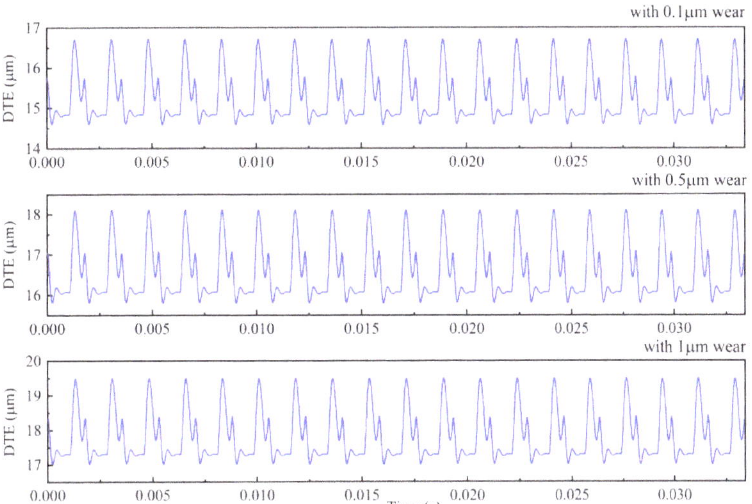

Figure 4. Dynamic transmission error (DTE) with different wear severities.

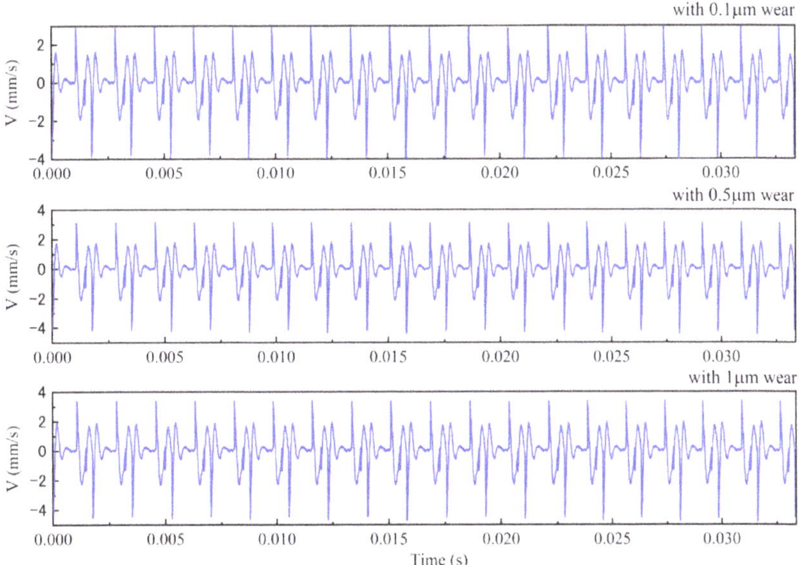

Figure 5. Vibrations of casing $\dot{y}_p$ under different wear severities.

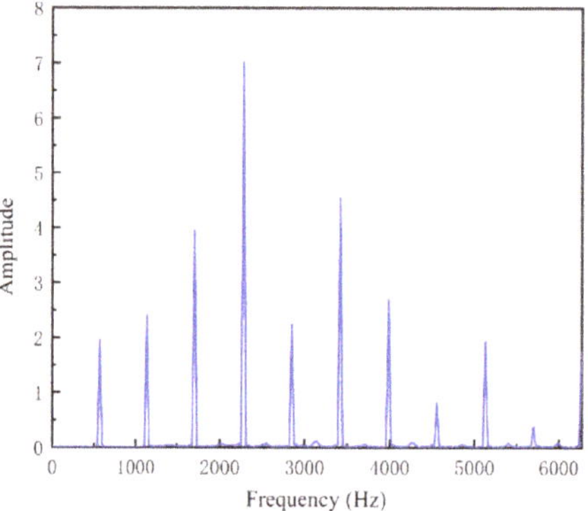

Figure 6. Frequency spectrum of $\dot{y}_p$ with 0.5 μm wear.

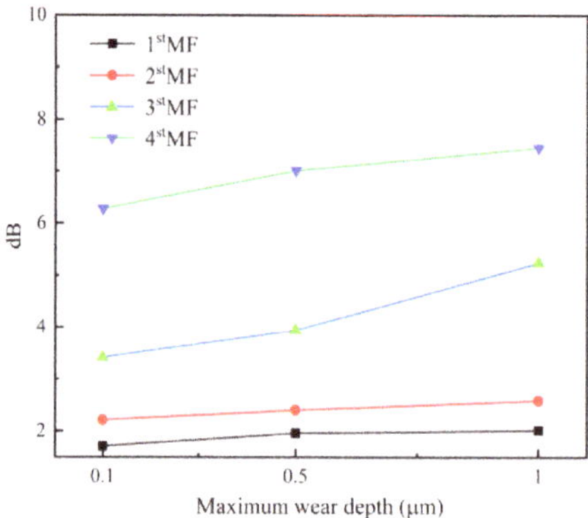

Figure 7. Spectral amplitudes of the first four meshing frequencies (MFs) alone with wear process.

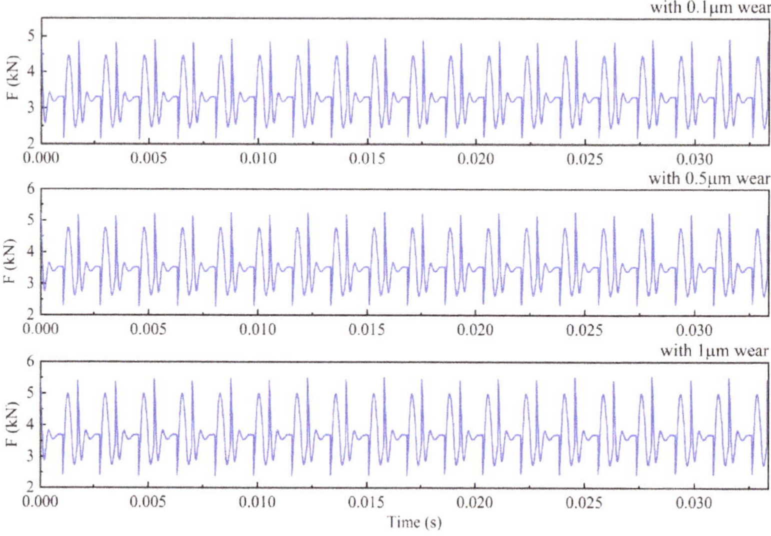

Figure 8. Gear mesh force $F^{(m)}$ of gear system under different wear severities.

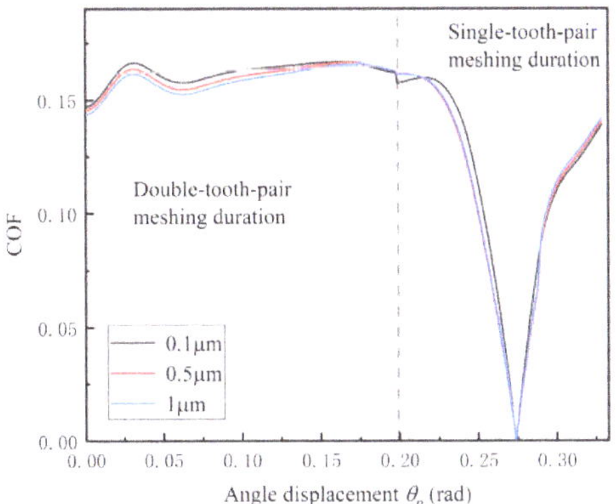

Figure 9. Friction coefficient under different wear severities.

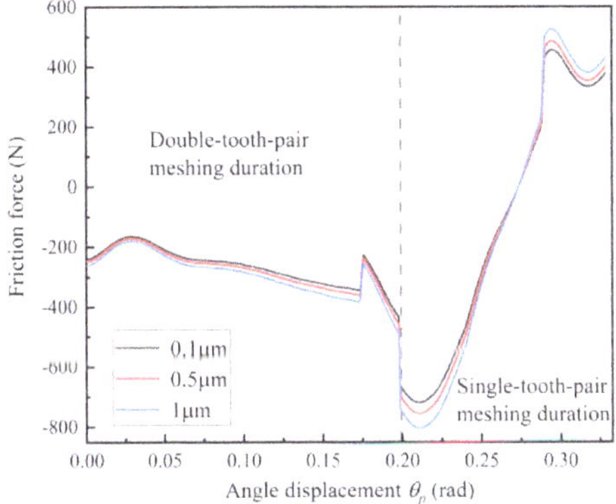

Figure 10. Friction force under different wear severities.

4.2. Effects of the Dynamic Response on Gear Surface Wear

To better understand the impact of gear dynamics on gear surface wear, the wear behaviors under quasi-static conditions are taken as a baseline for comparisons. In this case, the gear system is operating at a rotational speed of 30 Hz, while the torque of the brake remains as 60 Nm. In addition, the wear coefficient k_0 is set as 2.5×10^{-12} m^2/N, and the fixed wear cycle m is set as 1×10^4. During this period, the maximum accumulated wear depth at any point on the meshing surface reaches the predetermined wear threshold $\varepsilon^m = 0.5$ μm, the geometry of gear surface needs to be reconstituted. Then, the updated gear surfaces are fed into the dynamic model to renew the dynamic meshing force.

The comparisons of meshing force and wear depth distribution under static and dynamic conditions are illustrated in Figure 11. From Figure 11, it can be seen that the wear depth ΔH under dynamic conditions and static conditions are quite approximate in the initial stage. However, the wear depth ΔH under dynamic conditions become higher compared with static conditions with the wear process, as demonstrated in Figure 12. The reason is that dynamic contact force increases due to surface wear, as explained in Section 4.1, and the dynamic contact force is typically lager than the corresponding quasistatic force, ultimately, which results in a faster wear process. In consequence, the maximum accumulated wear depths under dynamic conditions exceed those under static conditions with the increase in wear cycle, as shown in Figure 13. Therefore, in order to guarantee accurate wear prediction results, the proposed dynamic wear prediction model is necessary.

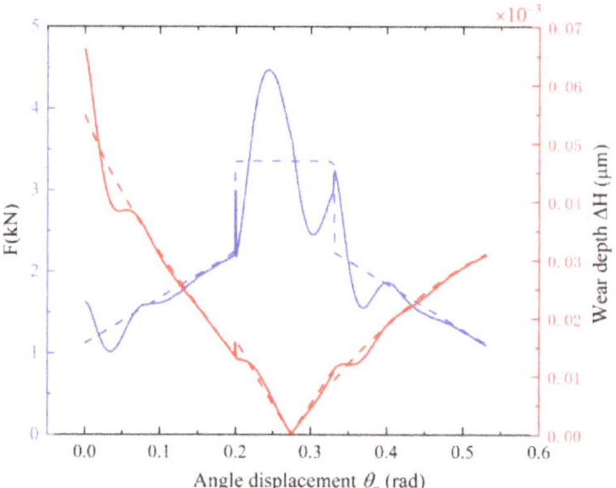

Figure 11. The meshing force and wear depth distribution of a pinion under static and dynamic conditions in the initial stage. (Dotted lines donate the static condition, and solid lines donate the dynamic condition).

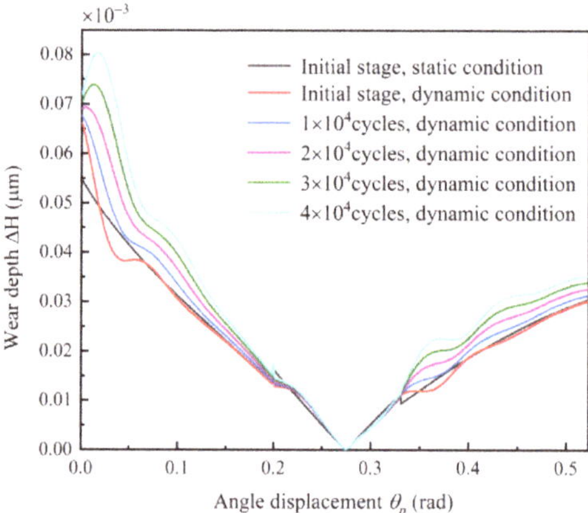

Figure 12. Evolutions of wear depth distribution of a pinion with wear process.

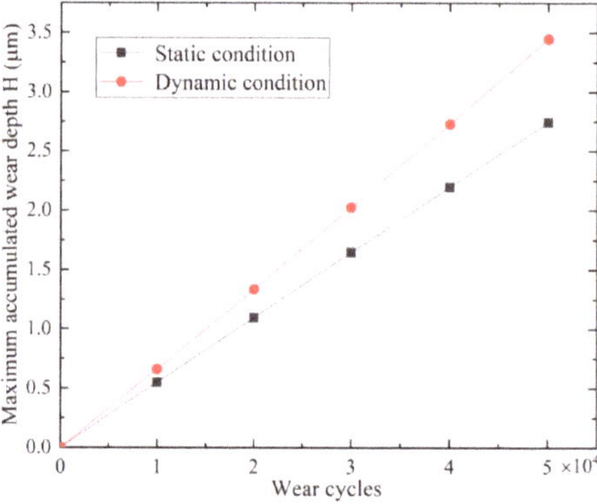

Figure 13. Maximum accumulated wear depth of a pinion under different conditions.

The proposed model has exhibited its capability in wear prediction. However, it also has some drawbacks. For example, it cannot consider the change in micro hardness caused by wear. Additionally, the wear coefficient is assumed as a constant value. However, the wear coefficient is susceptible to surface conditions, such as micro hardness and roughness [8,32]. In the future, the gear-wear prediction methodology proposed by Ke Feng et al. [33] which updated Archard's wear coefficient based on vibration responses will be applied to improve the present model. Moreover, though the proposed model has been verified through simulation analysis in this paper, experimental validations are lacking. In the future, the experiments under dry and lubricated conditions will be arranged to verify the proposed model.

5. Conclusions

In this paper, a dynamic model, which integrates TVMS, friction, and time-varying load-sharing ratio, is established, and then combine with the Archard wear model to establish a dynamic wear prediction model. An improved updated method of wear depth is proposed. With help of the proposed dynamic wear prediction model, the coupling effects between gear dynamic behaviors and gear surface wear process can be studied. Surface wear of a gear system results in the increase in dynamic responses, such as DTE, velocity, and meshing force. In turn, the increased meshing force results in faster wear process. Moreover, during the initial wear, the COF decrease slightly due to the reduction in surface roughness, but the friction force increases because of the improvement of dynamic meshing force. The ability of the proposed model in tracking and predicting wear process is superior to the classical static wear prediction model. Compared to the static prediction model, the proposed dynamic prediction model can update the wear depth distribution with wear process. Although the wear depth distributions under dynamic and static conditions are quite approximate in the initial stage, the maximum wear depth of a pinion under dynamic conditions already reaches 1.6 times that of the corresponding quasi-static condition. Therefore, the proposed dynamic wear prediction model is more reasonable for predicting the wear process.

Author Contributions: Conceptualization, J.R. and H.Y.; methodology, H.Y.; software, J.R.; validation, J.R. and H.Y.; formal analysis, J.R.; investigation, J.R.; resources, H.Y.; data curation, H.Y.; writing—original draft preparation, J.R.; writing—review and editing, J.R.; supervision, H.Y.; project administration, H.Y. All authors have read and agreed to the published version of the manuscript.

Funding: This research was funded by National Science and Technology Major Project of the Ministry of Science and Technology of China, grant number J2019-IV-0016-0084.

Institutional Review Board Statement: Not applicable.

Informed Consent Statement: Not applicable.

Data Availability Statement: Not applicable.

Conflicts of Interest: The authors declare no conflict of interest.

References

1. Fernandes, P.; McDuling, C. Surface contact fatigue failures in gears. *Eng. Fail. Anal.* **1997**, *4*, 99–107. [CrossRef]
2. Wojnarowski, J.; Onishchenko, V. Tooth wear effects on spur gear dynamics. *Mech. Mach. Theory* **2003**, *38*, 161–178. [CrossRef]
3. Osman, T.; Velex, P. Static and dynamic simulations of mild abrasive wear in wide-Faced solid spur and helical gears. *Mech. Mach. Theory* **2010**, *45*, 911–924. [CrossRef]
4. Jamari, J.; Ammarullah, M.I.; Santoso, G.; Sugiharto, S.; Supriyono, T.; van der Heide, E. In Silico Contact Pressure of Metal-On-Metal Total Hip Implant with Different Materials Subjected to Gait Loading. *Metals* **2022**, *12*, 1241. [CrossRef]
5. Choy, F.; Polyshchuk, V.; Zakrajsek, J.; Handschuh, R.; Townsend, D. Analysis of the effects of surface pitting and wear on the vibration of a gear transmission system. *Tribol. Int.* **1996**, *29*, 77–83. [CrossRef]
6. Kuang, J.H.; Lin, A.D. The effect of tooth wear on the vibration spectrum of a spur gear pair. *J. Vib. Acoust.* **2001**, *123*, 311–317. [CrossRef]
7. Yesilyurt, I.; Gu, F.; Ball, A.D. Gear tooth stiffness reduction measurement using modal analysis and its use in wear fault severity assessment of spur gears. *NDT E Int.* **2003**, *36*, 357–372. [CrossRef]
8. Yuksel, C.; Kahraman, A. Dynamic tooth loads of planetary gear sets having tooth profile wear. *Mech. Mach. Theory* **2004**, *39*, 695–715. [CrossRef]

9. Ding, H.; Kahraman, A. Interactions between nonlinear spur gear dynamics and surface wear. *J. Sound Vib.* **2007**, *307*, 662–679. [CrossRef]
10. Kahraman, A.; Ding, H. A Methodology to Predict Surface Wear of Planetary Gears Under Dynamic Conditions. *Mech. Based Des. Struct. Mach.* **2010**, *38*, 493–515. [CrossRef]
11. Liu, X.; Yang, Y.; Zhang, J. Investigation on coupling effects between surface wear and dynamics in a spur gear system. *Tribol. Int.* **2016**, *101*, 383–394. [CrossRef]
12. Flodin, A.; Andersson, S. Simulation of mild wear in spur gears. *Wear* **1997**, *207*, 16–23. [CrossRef]
13. Bajpai, P.; Kahraman, A.; Anderson, N.E. A Surface Wear Prediction Methodology for Parallel-Axis Gear Pairs. *J. Tribol.* **2004**, *126*, 597–605. [CrossRef]
14. Masjedi, M.; Khonsari, M. On the prediction of steady-State wear rate in spur gears. *Wear* **2015**, *342*, 234–243. [CrossRef]
15. Feng, K.; Borghesani, P.; Smith, W.A.; Randall, R.B.; Chin, Z.Y.; Ren, J.; Peng, Z. Vibration-Based updating of wear prediction for spur gears. *Wear* **2019**, *426*, 1410–1415. [CrossRef]
16. Feng, K.; Ji, J.; Ni, Q.; Beer, M. A review of vibration-Based gear wear monitoring and prediction techniques. *Mech. Syst. Signal Process.* **2023**, *182*, 109605. [CrossRef]
17. Mark, W.D. *Performance-Based Gear Metrology: Kinematic-Transmission-Error Computation and Diagnosis*; John Wiley & Sons: Hoboken, NJ, USA, 2012.
18. Jamari, J.; Ammarullah, M.; Afif, I.; Ismail, R.; Tauviqirrahman, M.; Bayuseno, A. Running-In Analysis of Transmission Gear. *Tribol. Ind.* **2021**, *43*, 434–441. [CrossRef]
19. He, S.; Gunda, R.; Singh, R. Effect of sliding friction on the dynamics of spur gear pair with realistic time-Varying stiffness. *J. Sound Vib.* **2007**, *301*, 927–949. [CrossRef]
20. Chen, S.; Tang, J.; Luo, C.; Wang, Q. Nonlinear dynamic characteristics of geared rotor bearing systems with dynamic backlash and friction. *Mech. Mach. Theory* **2011**, *46*, 466–478. [CrossRef]
21. Jiang, H.; Shao, Y.; Mechefske, C.K. Dynamic characteristics of helical gears under sliding friction with spalling defect. *Eng. Fail. Anal.* **2014**, *39*, 92–107. [CrossRef]
22. Luo, Y.; Baddour, N.; Liang, M. Dynamical modeling and experimental validation for tooth pitting and spalling in spur gears. *Mech. Syst. Signal Process.* **2018**, *119*, 155–181. [CrossRef]
23. Liu, P.; Zhu, L.; Gou, X.; Shi, J.; Jin, G. Modeling and analyzing of nonlinear dynamics for spur gear pair with pitch deviation under multi-State meshing. *Mech. Mach. Theory* **2021**, *163*, 104378. [CrossRef]
24. Pedrero, J.I.; Pleguezuelos, M.; Artés, M.; Antona, J.A. Load distribution model along the line of contact for involute external gears. *Mech. Mach. Theory* **2010**, *45*, 780–794. [CrossRef]
25. Xu, H.; Kahraman, A.; Anderson, N.E.; Maddock, D.G. Prediction of Mechanical Efficiency of Parallel-Axis Gear Pairs. *J. Mech. Des.* **2006**, *129*, 58–68. [CrossRef]
26. Liang, X.; Zuo, M.J.; Pandey, M. Analytically evaluating the influence of crack on the mesh stiffness of a planetary gear set. *Mech. Mach. Theory* **2014**, *76*, 20–38. [CrossRef]
27. Yang, D.C.H.; Lin, J.Y. Hertzian Damping, Tooth Friction and Bending Elasticity in Gear Impact Dynamics. *J. Mech. Transm. Autom. Des.* **1987**, *109*, 189–196. [CrossRef]
28. Yang, D.C.H.; Sun, Z.S. A Rotary Model for Spur Gear Dynamics. *J. Mech. Transm. Autom. Des.* **1985**, *107*, 529–535. [CrossRef]
29. Yu, W. Dynamic Modelling of Gear Transmission Systems with and without Localized Tooth Defects. Ph.D. Thesis, Queen's University, Kingston, ON, Canada, 2017.
30. Archard, J.F. Contact and Rubbing of Flat Surfaces. *J. Appl. Phys.* **1953**, *24*, 981–988. [CrossRef]
31. Brauer, J.; Andersson, S. Simulation of wear in gears with flank interference—A mixed FE and analytical approach. *Wear* **2003**, *254*, 1216–1232. [CrossRef]
32. Jamari, J.; Ammarullah, M.; Saad, A.; Syahrom, A.; Uddin, M.; van der Heide, E.; Basri, H. The Effect of Bottom Profile Dimples on the Femoral Head on Wear in Metal-On-Metal Total Hip Arthroplasty. *J. Funct. Biomater.* **2021**, *12*, 38. [CrossRef]
33. Feng, K.; Smith, W.A.; Peng, Z. Use of an improved vibration-Based updating methodology for gear wear prediction. *Eng. Fail. Anal.* **2020**, *120*, 105066. [CrossRef]

Article

A FCEEMD Energy Kurtosis Mean Filtering-Based Fault Feature Extraction Method

Chengjiang Zhou [1,2], **Ling Xing** [1,2,*], **Yunhua Jia** [1,2,*], **Shuyi Wan** [1,2,*] and **Zixuan Zhou** [1,2]

[1] The Laboratory of Pattern Recognition and Artificial Intelligence, Kunming 650500, China
[2] School of Information Science and Technology, Yunnan Normal University, Kunming 650500, China
* Correspondence: lin.xing@ynnu.edu.cn (L.X.); jia_yunhua@163.com (Y.J.); shuyi_wan@163.com (S.W.)

Abstract: Aiming at the problem that fault feature extraction is susceptible to background noises and burrs, we proposed a new feature extraction method based on a new decomposition method and an effective intrinsic mode function (IMF) selection method. Firstly, pairs of white noises with opposite signs were introduced to neutralize the residual noises in ensemble empirical mode decomposition (EEMD) and suppress mode mixing. Both the reconstruction error (1.8445×10^{-17}) and decomposition time (0.01 s) were greatly reduced through fast, complementary ensemble empirical mode decomposition (FCEEMD). Secondly, we integrated the energy and kurtosis of the IMF and proposed an effective IMF selection method based on energy kurtosis mean filtering, and the background noise of the signal was greatly suppressed. Finally, the periodic impacts were extracted from the IMF reconstruction signal by multipoint optimal minimum entropy deconvolution adjusted (MOMEDA). The fault frequencies were extracted from the periodic impacts through Hilbert demodulation, and the relative errors between the measured values and the theoretical values were all less than 0.05. The experimental results show that the proposed method can extract fault features more efficiently and provide a novel method for the fault diagnosis of rotating machinery.

Keywords: bearing; fast complementary ensemble empirical mode decomposition; energy kurtosis mean filtering; multipoint optimal minimum entropy deconvolution adjusted; feature extraction

Citation: Zhou, C.; Xing, L.; Jia, Y.; Wan, S.; Zhou, Z. A FCEEMD Energy Kurtosis Mean Filtering-Based Fault Feature Extraction Method. *Coatings* **2022**, *12*, 1337. https://doi.org/10.3390/coatings12091337

Academic Editor: Edoardo Proverbio

Received: 29 July 2022
Accepted: 8 September 2022
Published: 14 September 2022

Publisher's Note: MDPI stays neutral with regard to jurisdictional claims in published maps and institutional affiliations.

1. Introduction

Bearing is widely used in metallurgy, electric power and the machinery manufacturing industry [1]. Bearing failure affects the normal operation and production efficiency of equipment and even causes a serious loss of life and property [2]. Therefore, bearing fault diagnosis is important. The running process of the bearing is complicated, and the vibration signal is a nonlinear and non-stationary signal, which brings great challenges to fault diagnosis.

Hilbert envelope demodulation is the commonly used feature extraction method [3], which can identify the vibration shock contained in the modulated signals. Envelope demodulation is effective when used for a single frequency modulation and amplitude modulation (AM-FM) signal [4]. However, modulation signals are often mixed with carrier signals and noises. In order to obtain the AM-FM signal, the original signal needs to be decomposed before demodulation [5]. The commonly used decomposition algorithms include empirical mode decomposition (EMD) [6] and variational mode decomposition (VMD) [7], etc. The results of VMD depend on the setting of its two parameters. The intrinsic mode functions (IMFs) obtained by EMD have an endpoint effect problem and are mixed with each other. To solve the problems, ensemble empirical mode decomposition (EEMD) [8] and complementary ensemble empirical mode decomposition (CEEMD) [9] are proposed. Gao decomposed the bearing signal into several IMFs through EEMD, and selected the effective IMFs by the correlation coefficient and root mean square [10]. Although EEMD can suppress mode mixing to a certain extent, the added white noises cannot be eliminated completely [11]. In the CEEMD, white noises are added in pairs to the original signal to generate two sets of ensemble IMFs. CEEMD can eliminate the residue of added white noises totally no matter how many noises it contains [9]. Gu decomposed the bearing signal through CEEMD and selected the effective IMFs with a large correlation coefficient [11]. Li decomposed the underwater acoustic signals by complete ensemble empirical mode decomposition with adaptive noise (CEEMDAN) and identified noisy IMFs through the minimum mean square variance criterion [12]. Li decomposed warship radio noise signals by CEEMDAN and used a duffing chaotic oscillator to accurately detect the linear spectrum frequency of low-frequency IMFs [13], which can better suppress mode mixing. However, EEMD and CEEMD based on auxiliary noises have high computational complexity and low efficiency [14]. Wang proposed fast ensemble empirical mode decomposition (FEEMD) [15], which is gradually used in wind speed forecasting. Chegini decomposed the vibration signal into several IMFs by FEEMD and selected effective IMFs by autocorrelation function [16]. Sun determined the effective IMFs through FEEMD and correlation coefficient and improved the accuracy of wind speed forecasting [17]. The results show that the speed of FEEMD is fast, and the decomposition error is smaller than that of EEMD.

The IMFs obtained by EEMD and FEEMD contain carrier signals, impact components and noises, so it is important to select IMFs. Many scholars use energy, kurtosis and the correlation coefficient to select effective IMFs. If a selection method based on a single index is used, the IMFs related to the fault impact may be removed, and the IMFs related to noise may be retained. Xia decomposed the signal by VMD and selected effective IMFs with the maximum correlation coefficient and kurtosis [18]. Most of the noises can be removed by reconstructing the effective IMFs. However, there are still some noises in the reconstructed signal, which causes the impacts related to faults to be submerged in the noises. Therefore, the fault impacts need to be enhanced and extracted from the reconstructed signal, which can be regarded as a deconvolution process [19] and can be achieved by minimum entropy deconvolution (MED) and maximum correlation kurtosis deconvolution (MCKD), etc. He proposed an adaptive MED based on autocorrelation energy ratio, which makes it possible that recover a single random pulse when the filter length is not improper [20]. However, MED can only extract a single impact from the signal. Zhang selected effective IMFs through the weighted kurtosis index and then extracted the periodic impacts based on MCKD [21]. Cui decomposed the signal by VMD, finally selecting the sensitive mode through kurtosis and extracting periodic shocks [22] by MCKD. Although MCKD can extract a series of impacts, it cannot extract the complete periodic impacts [23]. Li decomposed the signal through the Hankel matrix and obtained various forms of noises, then extracted the periodic impacts of the reconstructed signal through multipoint optimal minimum entropy deconvolution adjusted (MOMEDA) [24]. Zhou decomposed the bearing signal through VMD and selected effective IMFs through the average of kurtosis, then extracted the periodic impacts through MOMEDA [25]. Xiao optimized the filter length of MOMEDA, then enhanced the periodic impacts and extracted the bearing fault frequencies [26]. Many studies have shown that the method combining MOMEDA with a decomposition algorithm can not only extract the periodic impacts contained in the noisy signal but also suppress the noise to a certain extent. In addition to the above methods, some very classic and useful methods are also worthy of our reference. Bayram [27] calculated the energy of the noisy and noise-free signal and obtained the wavelet coefficients that were used in classifying, proposing a feature extraction method based on the wavelet coefficient. Kaya [28] proposed a new feature extraction method based on co-occurrence matrices for bearing vibration signals, and the correlation, energy, homogeneity and contrast features were extracted from the co-occurrence matrices. To solve the spectral segmentation defect of the empirical wavelet transform (EWT) and improve its ability to extract fault features, Liu [29] proposed a new algorithm based on maximum envelope fitting and proposed a sensitive IMFs assessing index to obtain the characteristics of the time–frequency domain and amplitude domain to extract fault information. KAYA [30] proposed a technique based on continuous wavelet transform (CWT) and two dimensional (2D) convolutional neural networks (CNN) to detect the fault size from vibration data of various bearing failure types. Kuncan [31] proposed a bearing fault diagnosis method combined with one-dimensional local binary pattern and gray relational analysis, and the fault identification accuracy was shown to reach 99.3%. Han [32] extracted five feature vectors through freeman chain codes, then online diagnosis was achieved through an improved brainstorm optimization (BSO) algorithm.

However, some new problems can be found in fault feature extraction. Although FEEMD improves the decomposition efficiency, the white noises added to FEEMD cannot be eliminated completely, which leads to mode mixing. Secondly, the effective IMFs selection methods based on a single index cannot easily retain the IMFs related to vibration and shock, and the effect of noise suppression is also poor. Thirdly, the iterative of MED and MCKD is complicated, and the resulting filter is not necessarily a global optimal filter [24], which makes it difficult to extract periodic impacts effectively.

In view of the above problems, the contributions and innovations of the paper are as follows. To eliminate the mode mixing in FEEMD, pairs of white noises with opposite signs are added to the signal before each round of decomposition. To select the effective IMFs related to faults, the energy and kurtosis of each IMF are assigned a weight, respectively, and fused into an index, which is called the energy kurtosis weighted index. The average of all weighted indices is used as a threshold to select effective IMFs, called energy kurtosis mean filtering. The selected IMFs are reconstructed into a reconstructed signal. Finally, MOMEDA [23] is used to extract the periodic impacts in the reconstructed signal, and the fault frequencies can be extracted by Hilbert envelope demodulation. In summary, the proposed method can extract bearing fault features effectively, and it provides technical guidance for fault diagnosis.

The remainder of this paper is organized as follows. In Section 2, the methodology of the proposed feature extraction method is introduced in detail. In Section 3, the validity of the proposed method is verified through the Case Western Reserve University (CWRU) bearing datasets. In Section 4, the proposed method is applied to analyze the NASA bearing datasets, and the validity of the method is verified again. In Section 5, the superiority of the proposed method is validated by comparing various methods. The conclusions are summarized in Section 6.

2. The Proposed Feature Extraction Method

2.1. Fast Complementary Ensemble Empirical Mode Decomposition

FCEEMD is inspired by FEEMD [15] and CEEMD [9], which inherit the high efficiency of FEEMD and the low reconstruction error and weak mode mixing of CEEMD. If a set of signals is $X_t = [x_1, x_2, \ldots, x_n]$, the steps of FCEEMD are as follows.

(1) Initializing the amplitude of the added white noises, the number of ensemble trials I and the current ensemble trial number $i = 1$

(2) Pairs of white noises $\pm n_i(t), i = 1, 2, \ldots, n$ with opposite signs are added to the original signal $x(t)$ to generate two noise-added signals, $P_i(t)$ and $N_i(t)$.

$$\begin{cases} P_i(t) = x(t) + n_i(t) \\ N_i(t) = x(t) - n_i(t).' \end{cases} \tag{1}$$

where $x(t)$ represents the original signal and $n_i(t)$ represents white noises.

(3) Decompose the noise-added signals $P_i(t)$ and $N_i(t)$ into a series of IMFs.

$$P_i(t) = \sum_{j=1}^{q} c_{j,i}^1(t), \quad N_i(t) = \sum_{j=1}^{q} c_{j,i}^2(t)., \tag{2}$$

where $c_{j,i}^1(t)$ and $c_{j,i}^2(t)$ represent the j-th IMF in the i-th trial, and q represents the number of IMFs.

(4) If $i < I$, then repeat steps (2) and (3) with $i = i+1$, and add different white noises with opposite signs $\pm n_i(t), i = 1, 2, \ldots, n$ each time.

(5) Calculate the ensemble means of the $2I$ trials for each IMF as the results.

$$c_j(t) = \frac{1}{2I} \sum_{i=1}^{I} (c_{j,i}^1 + c_{j,i}^2).,$$

(3)

where $c_j(t)$ is the j-th IMF obtained by FCEEMD. To compare the decomposition performance of the EEMD-based methods under the same parameter conditions, the parameters of all methods are set according to the EEMD parameter criteria $\varepsilon_n = \varepsilon/\sqrt{N}$ [9], where N is the number of trials used to derive the ensemble IMFs, ε is the RMS amplitude of added noises and ε_n is the final standard deviation of error. To minimize ε_n, we refer to the parameter settings of CEEMD in references [11,33]. The amplitude of white noises is 0.2, and the number of ensemble trials is 50×2 [34]. In addition, the FCEEMD also has the following processing procedures.

(6) Stopping criteria in EEMD and CEEMD is replaced with the rule of fixed screening times, which reduces the quantities of computation while guaranteeing the EMD's second-order filter characteristics.

(7) In the process of spline interpolation, the popular Thomas algorithm, a simplified form of Gaussian elimination and the most efficient way to solve tridiagonal linear equations, is used to solve the tridiagonal matrices [35], which optimizes the time complexity and space complexity.

(8) In optimized programs, these files have file types of mexw32 or mexw64, which has the advantage of fast calculating speed.

FCEEMD has the same fundamentals as CEEMD, which can remove the residual white noises in FEEMD and EEMD by adding pairs of white noises $\pm n_i(t), i = 1, 2, \ldots, n$. It improves the efficiency of FCEEMD by optimizing the calculation procedure and program coding of the CEEMD; it can be applied to real-time signal processing and have increasingly extensive application prospects.

2.2. The Proposed Energy Kurtosis Mean Filtering

Suppose $\{x_1(t), x_2(t), \cdots x_N(t)\}$ is N IMFs obtained by FCEEMD, the energy E_i and kurtosis K_i of the i-th IMF are as follows.

$$E_i = \sum_{j=1}^{M} |x_{ij}(t)|^2.,$$

(4)

$$K_i = \frac{1}{M} \sum_{i=1}^{M} \left(\frac{x_{ij}(t) - \mu}{\sigma} \right)^4.,$$

(5)

where $x_{ij}(t)(i = 1, 2, \ldots, N, \ j = 1, 2, \ldots, M)$ represents the j-th data point in the i-th IMF, N represents the total number of IMFs and M is the data length. μ and σ represent the mean and standard deviation of $x_i(t)$, respectively. The energy E_i and kurtosis K_i of the i-th IMF can be calculated by the formulas above.

The energy can characterize the signal strength of different frequency bands. Kurtosis is independent of bearing speed, size and load, and it is sensitive to the impact signal. Different types of faults cause different changes in energy and kurtosis in different frequency bands. If only energy is used as the selection criterion for effective IMFs, a series of IMFs related faults may be ignored. If only kurtosis is used as the selection criterion for effective IMFs, the energy information of the signal may be lost. Therefore, the signal characteristics need to be considered comprehensively when selecting the IMFs. For the energy sequence $\{E_1, E_2, \ldots, E_N\}$ and the kurtosis sequence $\{K_1, K_2, \ldots, K_N\}$ obtained above, we introduced an effective weighting method to integrate them. The obtained index is called the energy kurtosis weighted index.

(1)   The regularized energy sequence E_{ni} and kurtosis sequence K_{ni} can be obtained by Formulas (6) and (7).

$$E_{ni} = (E_i - \min_{i=1}^{N} E_i) / (\max_{i=1}^{N} E_i - \min_{i=1}^{N} E_i)., \tag{6}$$

$$K_{ni} = (K_i - \min_{i=1}^{N} K_i) / (\max_{i=1}^{N} K_i - \min_{i=1}^{N} K_i)., \tag{7}$$

where N represents the total number of IMFs, and E_i and K_i are the energy and kurtosis of the i-th IMF, respectively.

(2)   The mean μ_e and mean square error σ_e of E_{ni} are calculated by Formulas (8) and (9), and the mean μ_k and mean square error σ_k of K_{ni} are calculated by Formulas (10) and (11).

$$\mu_e = \frac{1}{N} \sum_{i=1}^{N} E_{ni}., \tag{8}$$

$$\sigma_e = \sqrt{\frac{1}{N} \sum_{i=1}^{N} (E_{ni} - u_e)^2}., \tag{9}$$

$$\mu_k = \frac{1}{N} \sum_{i=1}^{N} K_{ni}. \tag{10}$$

$$\sigma_k = \sqrt{\frac{1}{N} \sum_{i=1}^{N} (K_{ni} - u_k)^2}., \tag{11}$$

where N represents the number of IMFs, and μ_e, σ_e, μ_k and σ_k can be obtained by the formulas.

(3)   The energy weight coefficient W_e is calculated by Formula (12), and the kurtosis weight coefficient is $W_k = 1 - W_e$. The energy kurtosis weighted index EK_i and its mean λ_{EK} can be calculated as follows.

$$W_e = \frac{\sigma_e}{\sigma_e + \sigma_k}., \tag{12}$$

$$EK_i = W_e E_{ni} + W_k K_{ni}., \tag{13}$$

$$\lambda_{EK} = \frac{1}{N} \sum_{i=1}^{N} EK_i., \tag{14}$$

where E_{ni} and K_{ni} are the regularized energy and kurtosis sequences of the i-th IMF, respectively, and N represents the number of IMFs.

The more obvious the fault impacts contained in the IMF, the greater the energy kurtosis weighted index. We calculated the mean of EK_i and took it as a threshold. If the energy kurtosis weighted index of an IMF is smaller than the threshold, it is considered that the IMF does not contain a stable impact component and will be removed; otherwise, the IMF is selected as an effective IMF.

The reconstructed signal can be obtained by bit-by-bit accumulation of each effective IMF. Suppose x is the reconstructed signal, y is the impact component and h is the transfer function and e is the noise, then,

$$x = h * y + e. \tag{15}$$

The fault impact y can be extracted from x, and the process can be regarded as a deconvolution process [19]. MOMEDA [23] is a deconvolution algorithm, and its purpose is to find the optimal FIR filter in a non-iterative way and reset y through x. *Multi D-Norm* is proposed to deconvolute multiple impulses at a determined location, and its maximum problem is as follows.

$$Multi \quad D - Norm = MDN(\vec{y}, \vec{t}) = \frac{1}{\left\|\vec{t}\right\| \left\|\vec{y}\right\|} \frac{\vec{t}^T \vec{y}}{}, \tag{16}$$

$$MOMEDA = \max_{\vec{f}} MDN(\vec{y}, \vec{t}) = \max_{\vec{f}} \frac{\vec{t}^T \vec{y}}{\left\|\vec{y}\right\|}, \tag{17}$$

where $\vec{t}$ represents a constant vector, and it describes the position and weight of the target impact. $\vec{y}$ represents impact vector. The optimal filter can be obtained by solving the maximum of the D-norm.

To measure the position of the fault-related maximum pulse point accurately, Multipoint Kurtosis is introduced.

$$MKurt = \left(\sum_{n=1}^{N-L} t_n^2\right)^2 \sum_{n=1}^{N-L} (t_n y_n)^4 / \left(\sum_{n=1}^{N-L} t_n^8 \left(\sum_{n=1}^{N-L} y_n^2\right)^2\right). \tag{18}$$

To compare the results, the default parameters of MOMEDA are considered, the filter length is 500 ($L = 500$), and the period range is $T = [10 : 0.1 : 300]$. More information about MOMEDA can be found in ref. [23]. After obtaining the fault impacts, the Hilbert envelope demodulation can extract the fault features more easily and effectively.

2.3. The FCEEMD Energy Kurtosis Mean Filtering Based Fault Feature Extraction Method

Aiming at the defects of the existing methods, a FCEEMD energy kurtosis mean filtering-based fault feature extraction method is proposed, and the method is used to analyze the CWRU and NASA datasets, as shown in Figure 1.

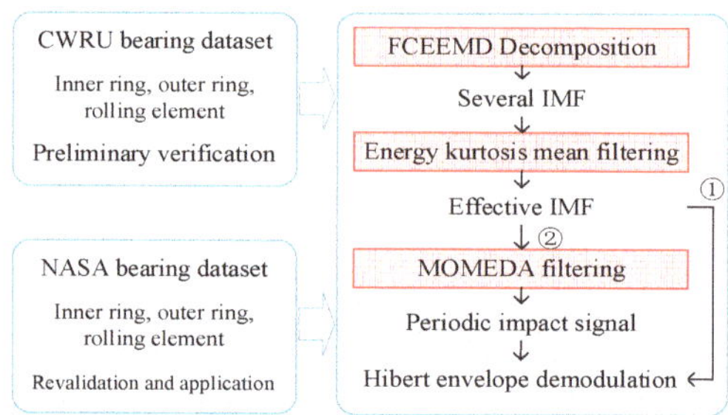

Figure 1. The experimental verification of the proposed method.

(1) To suppress mode mixing and improve the decomposition efficiency, the proposed FCEEMD is used to decompose bearing signals into several IMFs.

(2) To select the effective IMFs related to the fault impact and reduce the reconstruction error, the energy kurtosis weighted index of each IMF is calculated. If the energy kurtosis weighted index of an IMF is greater than the mean of all indexes, it indicates that the IMF contains the main information of failure, so it is chosen as an effective IMF. All effective IMFs can be obtained this way, and their cumulative sum is used as the reconstructed signal.

(3) To suppress the interference frequencies and extract the fault frequencies effectively, the periodic impacts in the reconstructed signal are extracted by MOMEDA, and the fault frequencies are extracted by Hilbert envelope demodulation.

Firstly, the inner ring, outer ring and rolling element signals of CWRU bearing are analyzed to verify the validity of the proposed method, as shown in Figure 1.

Secondly, to verify the universality and application ability of the proposed method, the method is used to extract the fault features of the inner ring, outer ring and rolling element signals in the NASA dataset, as shown in Figure 1.

Finally, to verify the superiority of the proposed method, the method is compared with a variety of feature extraction methods in the case of rolling element feature extraction, as shown in Figure 2.

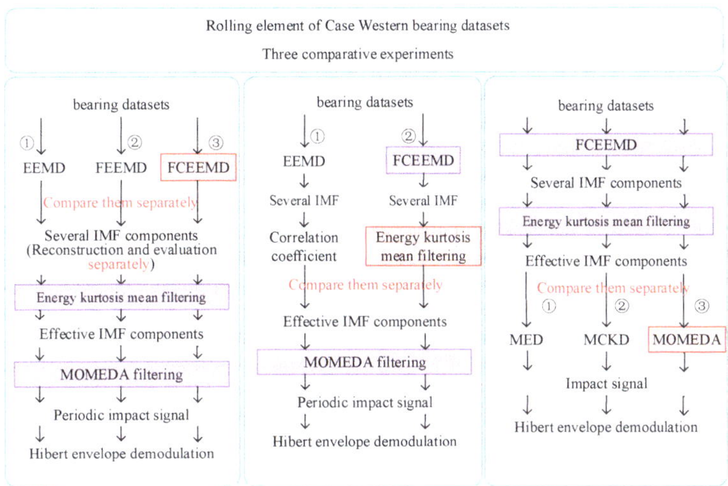

Figure 2. Comparison of feature extraction methods.

(1) To verify the superiority of the FCEEMD-based method, the rolling element signal is decomposed by EEMD, FEEMD and FCEEMD, and the decomposition results and feature extraction effects are compared and evaluated.

(2) To verify the superiority of FCEEMD energy kurtosis mean filtering, the extraction results of this method are compared with those of the method in [36].

(3) To verify the superiority of the MOMEDA-based method, the proposed method is compared with the MED and MCKD-based methods.

3. Experimental Verification Based on CWRU Bearing Data

To verify the validity of the proposed method, the CWRU dataset [36] is analyzed. The specification of the bearing is 6205-2RSJEMSKF, the motor speed is 1797 rpm, and the rotational frequency is 1797/60 Hz = 29.95 Hz. The parameters are shown in Table 1.

Table 1. The structure parameters of SKF 6205 bearing.

Ball Diameter (d)	Number of Balls (N)	Pitch Diameter (D)	Contact Angle (α)
7.938 mm	9	39 mm	0

Sensors are installed on the drive motor, and the vibration signals under different damage diameters are collected. The sampling frequency is 12 kHz, and the experimental data length is 2048. The signals of the inner ring, outer ring and rolling element with a damage diameter of 0.007 inches are used for experimental analysis. According to the parameters and the formulas in refs. [37,38], the fault frequencies can be obtained as shown in Table 2. The experimental environment is Intel (R) Core (TM) i5-5200u CPU @ 2.20 GHz.

Table 2. Fault characteristic frequency (unit: Hz).

Inner Ring Fault	Outer Ring Fault	Rolling Element Failure
162.1852	107.3648	141.1693

3.1. Inner Ring Fault Feature Extraction

The time and frequency domain waveforms of the inner ring fault signal are shown in Figure 3. Due to the influence of noises and interferences, it is difficult to extract fault frequency directly from the frequency domain waveform. Therefore, the inner ring signal is decomposed by FCEEMD, and the effective IMFs are selected by energy kurtosis mean filtering. In all subsequent experiments, all parameters of the FCEEMD algorithm are consistent, where the number of IMFs is 9, the number of screening iterations maxSift is 10 and the rest of the parameters are default values.

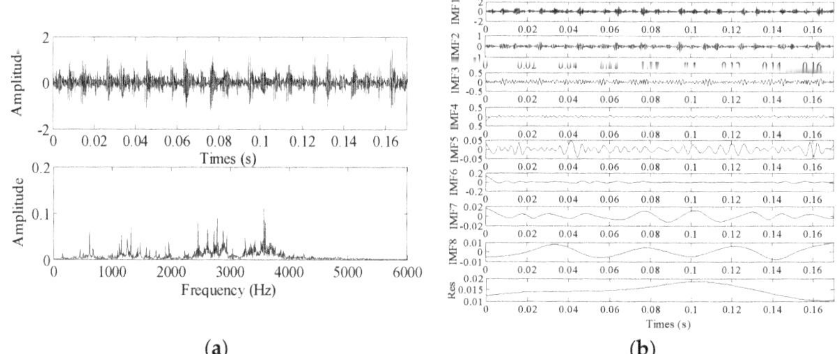

(a)   (b)

Figure 3. CWRU inner ring fault signal and its decomposition results. (**a**) Time and frequency domain analysis of inner ring signal. (**b**) FCEEMD decomposition result of inner ring fault signal.

As shown in Figure 4a, the energy of the IMFs decreases gradually, the IMF1 contains most of the energy and the change in the kurtosis has no obvious regularity. According to the proposed method, the IMFs whose energy kurtosis weighted index is less than the threshold are regarded as noises and removed. Therefore, the effective IMFs include IMF1, IMF4 and IMF6, and the envelope demodulation of the effective IMFs reconstruction signal is shown in Figure 4b.

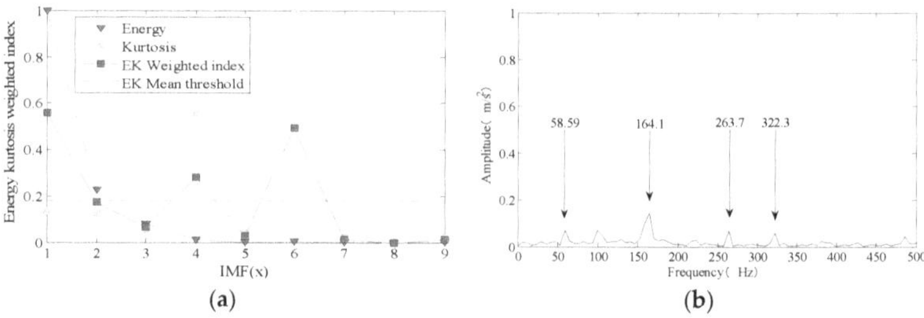

(a)   (b)

Figure 4. The result of energy kurtosis means filtering. (**a**) Energy kurtosis weighted index of the CWRU inner ring IMFs. (**b**) Envelope demodulation of CWRU inner ring reconstruction signal.

As shown in Figure 4b, the envelope spectrum shows an obvious peak at 164.1 Hz, and the peak frequency is close to the theoretical value of 162.1852 Hz of fault frequency of the inner ring. Meanwhile, the peak frequency 322.3 Hz is close to the double frequency 324.3704 Hz, and the peak frequency 58.59 Hz is close to the double frequency 59.9 Hz of the rotating frequency. Due to the influence of noises and interferences, the absolute error between the measured value and theoretical values of the inner ring fault frequency is 164.1 Hz − 162.1852 Hz = 1.9148 Hz, and the relative error is 0.0118. The results show that the method based on FCEEMD energy kurtosis mean filtering is effective. However, there are still many peaks at 263.7 Hz and other unrelated frequencies, and these peaks have an impact on the results. Therefore, we extract the periodic impacts from the effective IMFs reconstruction signal through MOMEDA, and the results are shown in Figure 5a,b. In MOMEDA, the filter size L is 500, the window function is a rectangular window (ones (1,1)) and the remaining parameters are default parameters used in all subsequent experiments.

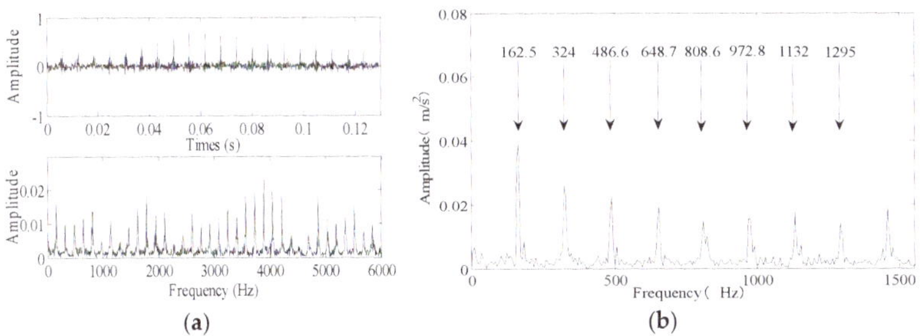

(a) (b)

Figure 5. MOMEDA filtered output of CWRU inner ring signal and its envelope demodulation. (a) Time and frequency domain analysis of MOMEDA filtered output for inner ring signal. (b) Envelope demodulation of MOMEDA filtered output for inner ring signal.

As shown in Figure 5b, the MOMEDA extracts the periodic impacts in the effective IMFs reconstructed signal and weakens the frequencies unrelated to the fault impact. It is noteworthy that the peaks appear at 162.5 Hz, 324 Hz, 486.6 Hz and 648.7 Hz, etc. These peaks correspond to the inner ring fault frequency 162.1852 Hz and its double frequency 324.3704 Hz, triple frequency 486.5556 Hz, quadruple frequency 648.7408 Hz and so on. After adding MOMEDA, the absolute error between the measured value and theoretical value is 162.5 Hz − 162.1852 Hz = 0.3148 Hz, and the relative error is 0.0019, which greatly reduces the frequency error and improves the fault detection accuracy.

3.2. Outer Ring Fault Feature Extraction

The time and frequency domain waveforms of the outer ring fault signal are shown in Figure 6a. The fault frequency is submerged by the complex frequency components. Therefore, the outer ring signal is decomposed by FCEEMD, and the effective IMFs are selected by energy kurtosis mean filtering.

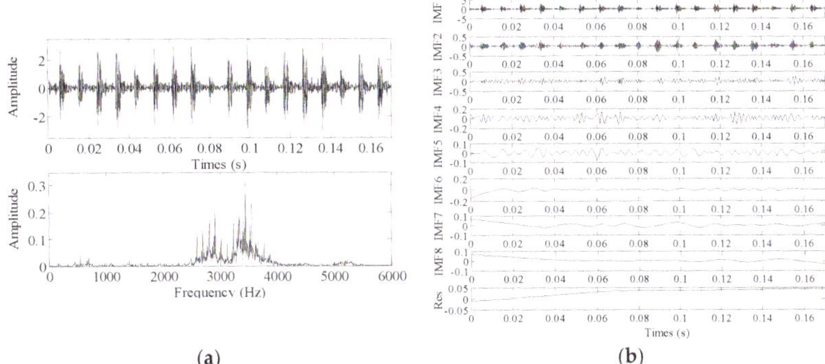

(a) (b)

Figure 6. CWRU outer ring fault signal and its decomposition results. (a) Time and frequency domain analysis of outer ring signal. (b) FCEEMD decomposition result of outer ring fault signal.

As shown in Figure 7a, the energy of the IMFs decreases gradually, and IMF1 contains most of the energy. The kurtosis of IMF6 is the largest, and the kurtosis of other IMF shows no regularity in variation. According to energy kurtosis mean filtering, the effective IMFs include IMF1 and IMF6, and the envelope demodulation of the reconstruction signal is shown in Figure 7b.

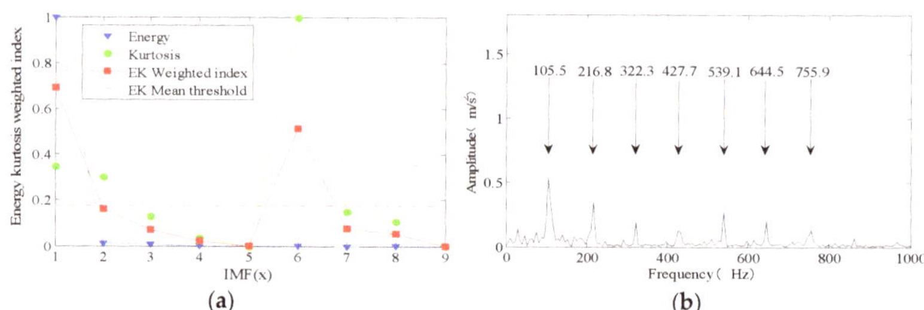

Figure 7. The result of energy kurtosis mean filtering. (**a**) Energy kurtosis weighted index of the CWRU outer ring IMF components. (**b**) Envelope demodulation of CWRU outer ring reconstruction signal.

The outer ring fault can be accurately diagnosed by combining the FCEEMD with energy kurtosis mean filtering. It is noteworthy that the peaks appear at 105.5 Hz, 216.8 Hz, 322.3 Hz and 427.7 Hz, etc. These peaks correspond to the outer ring fault frequency 107.3648 Hz and its double frequency 214.7296 Hz, triple frequency 322.0944 Hz, quadruple frequency 429.4592 Hz and so on. The results show that the method based on FCEEMD energy kurtosis mean filtering is extremely effective. Due to the influence of the parameter error and transmission path, the absolute error between the measured value and theoretical value is 107.3648 Hz − 105.5 Hz = 1.8648 Hz, and the relative error is 0.0174. Therefore, we extract the periodic impacts from the reconstruction signal through MOMEDA. The results are shown in Figure 8a,b.

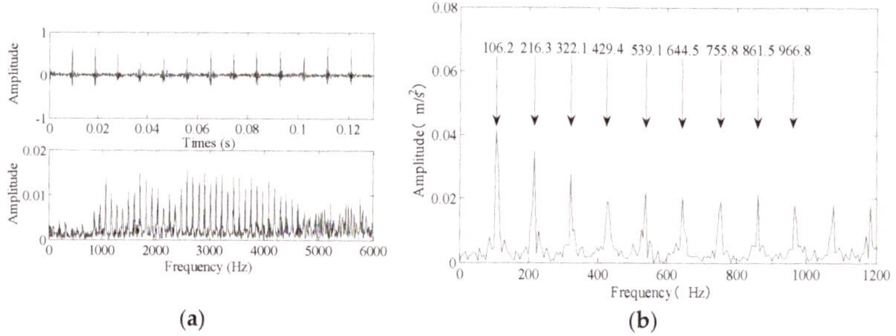

Figure 8. MOMEDA filtered output of CWRU outer ring signal and its envelope demodulation. (**a**) Time and frequency domain analysis of MOMEDA filtered output for outer ring signal. (**b**) Envelope demodulation of MOMEDA filtered output for outer ring signal.

MOMEDA effectively extracts the periodic impacts in the effective IMFs reconstructed signal and weakens the frequencies unrelated to the fault impacts. The absolute error between the measured value and theoretical value is 107.3648 Hz − 106.2 Hz = 1.1648 Hz, and the relative error is 0.0108. Compared with the results before MOMEDA filtering, MOMEDA reduces the deviation between the measured value and the theoretical value of the fault frequency. The outer ring fault frequency harmonics are obvious, and the outer ring fault can be diagnosed accurately. The results show that the proposed method is quite lightweight and efficient, and the precision of fault recognition is therefore improved.

3.3. Rolling Element Fault Feature Extraction

The time and frequency domain waveforms of the rolling element signal are shown in Figure 9a. Compared with the inner ring and the outer ring signals, the periodic impacts of the rolling element signal are not obvious. Moreover, the fault frequency is submerged by complex frequency components, and it is difficult to extract fault frequency from the frequency domain waveform. Therefore, the rolling element signal is decomposed by FCEEMD, and the effective IMFs are selected by energy kurtosis mean filtering.

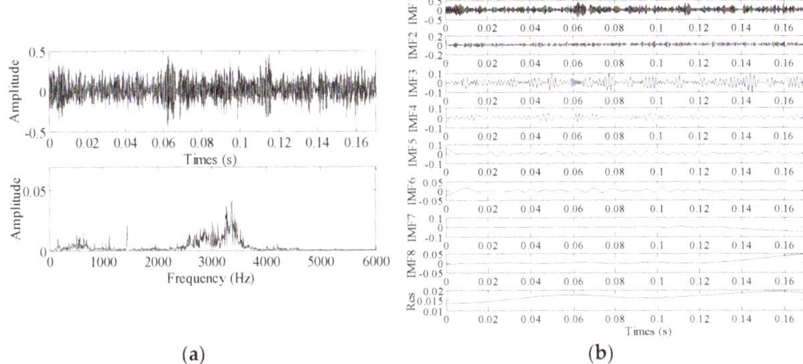

(a)   (b)

Figure 9. CWRU rolling element fault signal and its decomposition results. (**a**) Time and frequency domain analysis of rolling element signal. (**b**) FCEEMD decomposition result of rolling element fault signal.

As shown in Figure 10a, the energy of the IMFs decreases gradually; the IMF1 contains most of the energy. The kurtosis of the IMF increases gradually; the kurtosis of IMF6 is the largest, and then the kurtosis decreases gradually. If only the effective IMFs are selected based on energy, a lot of impact information will be lost. The selection of effective IMFs based on a single index will lead to incomplete fault information. It also shows that it is necessary to select IMF by energy kurtosis mean filtering. The effective IMFs include IMF1, IMF4, IMF5, IMF6, IMF7 and IMF8. The envelope demodulation of the reconstruction signal is shown in Figure 10b.

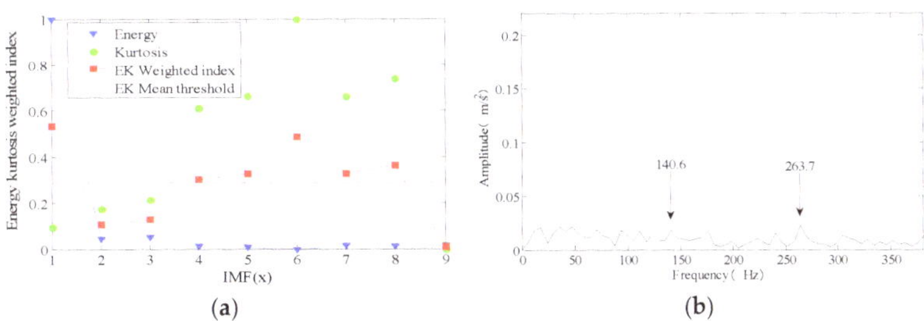

(a)   (b)

Figure 10. The result of energy kurtosis means filtering. (**a**) Energy kurtosis weighted index of the CWRU rolling element IMF components. (**b**) Envelope demodulation of CWRU rolling element reconstruction signal.

As shown in Figure 10b, the envelope spectrum shows a peak at 140.6 Hz, which is close to the rolling element fault frequency of 141.1693 Hz. However, there is a large deviation between the peak frequency 263.7 Hz and the double frequency 282.3386, and many irrelevant frequencies appear in the envelope spectrum. The non-stationary components contained in the rolling element signal are numerous and complex. These irrelevant components make it extremely difficult to extract fault features. We extracted the periodic impacts from effective IMFs reconstruction signal through MOMEDA, and the results are shown in Figure 11a,b.

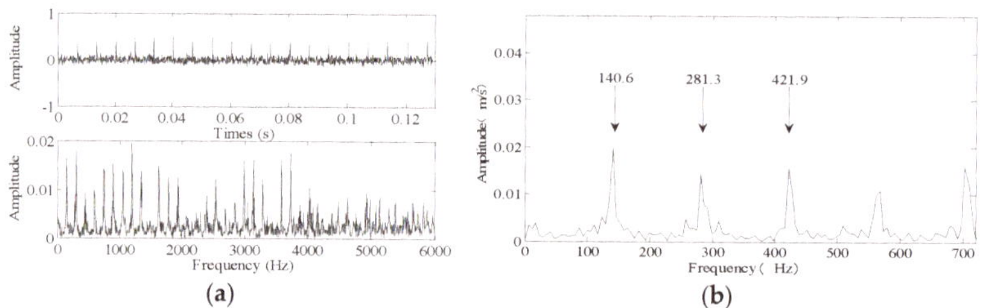

Figure 11. MOMEDA filtered output of CWRU rolling element signal and its envelope demodulation. (a) Time and frequency domain analysis of MOMEDA filtered output for rolling element signal. (b) Envelope demodulation of MOMEDA filtered output for rolling element signal.

In Figure 11b, although the impact characteristics of the rolling element signal are not obvious, MOMEDA can still extract the periodic impacts in the reconstructed signal effectively and weaken the frequencies unrelated to the faults. It is noteworthy that the peaks appear at 140.6 Hz, 281.3 Hz and 421.9 Hz, etc. These peaks correspond to the rolling element fault frequency 141.1693 Hz and its double frequency 282.3386 Hz, triple frequency 423.5079 Hz and so on. Due to the influence of noises and interferences, there are some deviations between the measured value and the theoretical value.

4. Method Verification and Application

To verify the reliability of the proposed method, the NASA bearing dataset is analyzed. The experimental platform [39] is shown in Figure 12a, and four Rexnord ZA-2115 bearings are mounted on the same spindle. The rotational speed of the spindle is 2000 rpm in the motor drive, and the rotating frequency (Fs) of the bearing is 33.33 Hz (fr = 2000/60 = 33.33 Hz). The PCB353B33 sensors are installed on the bearing housing, and the signal is collected by the DAQ-6062E acquisition card. The sampling frequency is 20 kHz, and the length of each set of data is 20480. The inner ring signal from the Bearing 3 of the No.1 dataset, the outer ring signal from the Bearing 1 of the No.2 dataset, the rolling element signal from Bearing 3 of the NO.1 dataset and the data length is 2048.

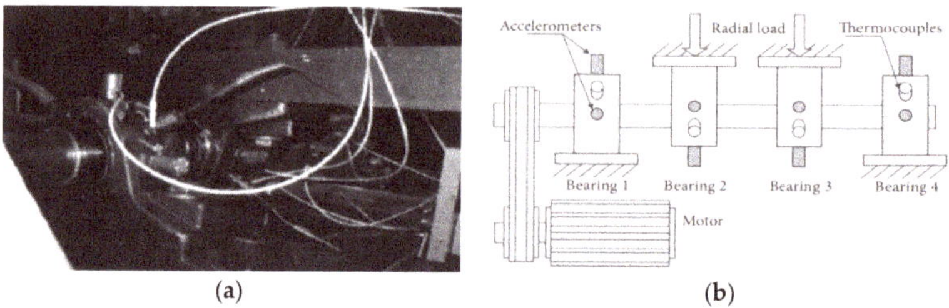

Figure 12. Illustration of the bearing experiment platform. (a) Bearing experiment platform. (b) Sensor layout.

According to the formulas of fault frequency in ref. [38] and the parameters in Table 3, the fault frequencies of NASA bearing are shown in Table 4. The important frequencies mainly include the ball pass frequency of the inner race (BPFI), the ball pass frequency of the outer race (BPFO), the ball spin frequency (BSF) and rotating frequency (Fs).

Table 3. The bearing structure factor of Rexnord ZA-2115.

Ball Diameter (d)	Number of Balls (Z)	Pitch Diameter (D)	Contact Angle (α)
8.407 mm	16	71.501 mm	15.17

Table 4. Fault characteristic frequency (unit: Hz).

Inner Ring Fault (BPFI)	Outer Ring Fault (BPFO)	Rolling Element Failure (BSF)
296.91	236.4	139.9

4.1. Inner Ring Fault Feature Extraction Application

The time and frequency domain waveforms of the inner ring signal are shown in Figure 13a. The signal contains some impact components. However, the period of these impact signals is not significant compared with the inner ring signal of the CWRU dataset. The fault frequency of the inner ring is submerged by a series of complex frequencies, which makes it particularly difficult to extract the fault feature. Therefore, the signal is decomposed by FCEEMD, and the effective IMFs are selected by energy kurtosis mean filtering.

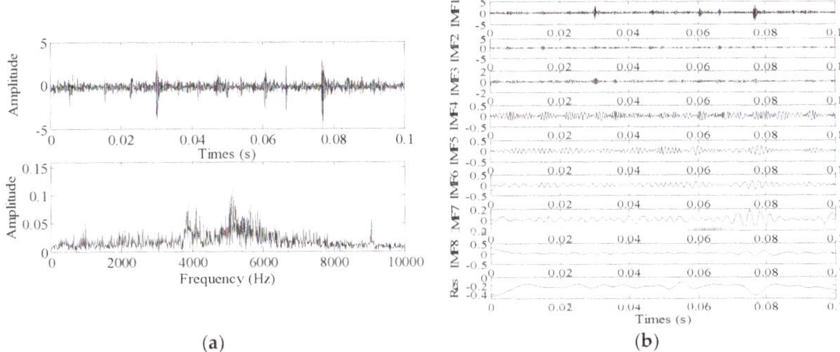

Figure 13. NASA inner ring fault signal and its decomposition results. (**a**) Time and frequency domain analysis of inner ring signal. (**b**) FCEEMD decomposition result of inner ring fault signal.

As shown in Figure 14a, the energy of the IMFs decreases gradually, and IMF1 contains most of the energy. The kurtosis of the IMF3 is the largest, and the kurtosis of other IMFs shows no regularity in variation. The IMFs whose energy kurtosis weighted index is less than the threshold are regarded as false components and removed. Therefore, the effective IMFs include IMF1, IMF2 and IMF3, and the envelope demodulation of the effective IMF reconstruction signal is shown in Figure 14b.

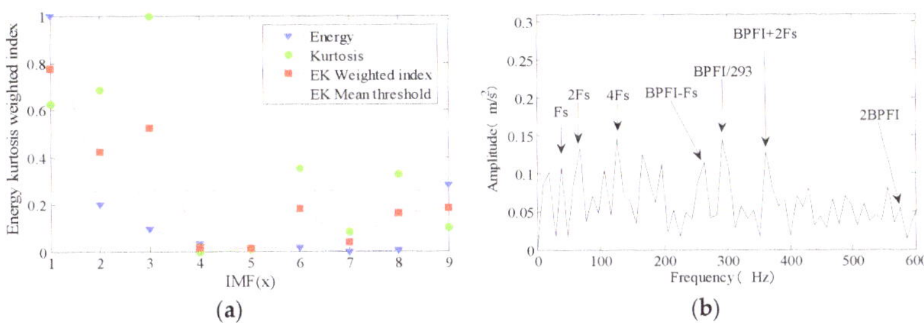

Figure 14. The result of energy kurtosis means filtering. (**a**) Energy kurtosis weighted index of the NASA inner ring IMF components. (**b**) Envelope demodulation of NASA inner ring reconstruction signal.

As shown in Figure 14b, the envelope spectrum shows an obvious peak at 293 Hz, which is close to the fault frequency of 296.9035 Hz. Meanwhile, the rotating frequency Fs = 33.33 Hz and its frequency multiplication appear in the low-frequency part. The modulation frequencies of inner ring fault frequency (BSFI) and rotating frequency (Fs) also appear in the envelope spectrum. Because of the influence of noises, there is some deviation between the theoretical and the measured value. The experimental results show the effectiveness of the method based on FCEEMD energy kurtosis mean filtering. However, the envelope spectrum still contains frequencies unrelated to the faults. These frequencies interfere with the identification of the fault frequency. We extract the periodic impacts from the reconstructed signal through MOMEDA.

As shown in Figure 15b, MOMEDA effectively extracts the periodic impacts in the reconstructed signal and weakens the frequencies unrelated to the faults. It is noteworthy that peaks appear at 293 Hz, 585.9 Hz, 878.9 Hz and 1172 Hz, etc. These peaks correspond to the inner ring fault frequency 296.91 Hz and its double frequency 593.82 Hz, triple frequency 890.73 Hz, quadrupled frequency 1187.64 Hz and so on. Because of the influence of noises, there is some deviation between the theoretical value and the measured value. The inner ring fault can be identified accurately according to the characteristic frequency.

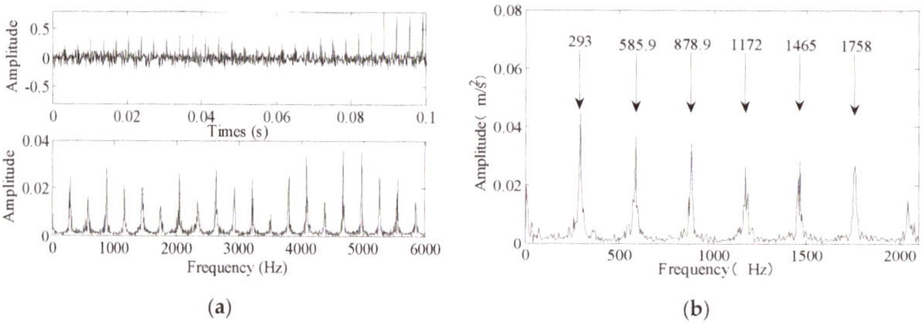

Figure 15. MOMEDA filtered output of NASA inner ring signal and its envelope demodulation. (**a**) Time and frequency domain analysis of MOMEDA filtered output for inner ring signal. (**b**) Envelope demodulation of MOMEDA filtered output for inner ring signal.

4.2. Outer Ring Fault Feature Extraction Application

The time and frequency domain waveforms of the outer ring signal are shown in Figure 16a. There are many interference frequencies in the frequency domain diagram, and the fault frequency is not obvious. Therefore, the outer ring signal is decomposed by FCEEMD, and the effective IMFs are selected by energy kurtosis mean filtering.

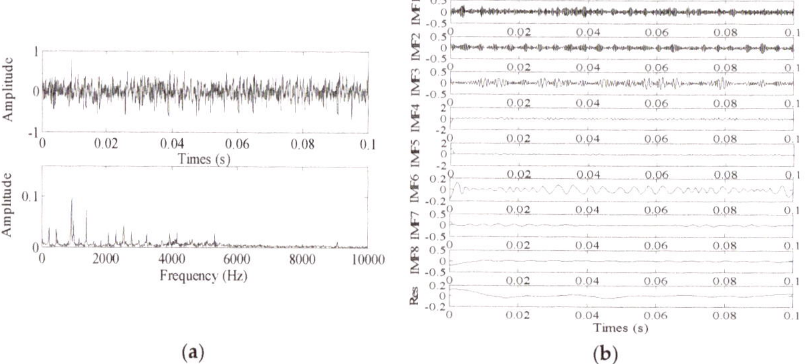

Figure 16. NASA outer ring fault signal and its decomposition results. (**a**) Time and frequency domain analysis of outer ring signal. (**b**) FCEEMD decomposition result of outer ring fault signal.

As shown in Figure 17a, the changing trend of the energy and kurtosis of the IMFs is irregular. According to the proposed method, the effective IMFs include IMF4 and IMF5, and the envelope demodulation of the effective IMFs reconstruction signal is shown in Figure 17b. Two obvious spectrum peaks appear at 61.59 Hz and 459 Hz. The two peaks correspond to the double frequency of the rotating frequency (2Fs = 66.66 Hz) and the double frequency of the outer ring fault frequency (472.8 Hz). The envelope spectrum shows three obvious peaks in the low-frequency part, and the three peaks correspond to the rotating frequency 33.33 Hz and its double frequency 66.66 Hz, with frequency quintupling at 166.65 Hz. Moreover, a peak appears at 224.6 Hz, which is close to the fault frequency of 236.4 Hz of the outer ring. The modulation frequencies of the outer ring fault frequency (BSFO) and rotating frequency (Fs) also appear in the envelope spectrum. Because of the influence of noises, the absolute error between the measured value and theoretical value is 236.4 Hz − 224.6 Hz = 11.8 Hz, and the relative error is 0.0499. Although the envelope spectrum contains clear fault frequencies, there are still many interference frequencies, and the fault features are not easy to be extracted. We extract the periodic impacts from the reconstructed signal through MOMEDA. The results are shown in Figure 18a,b.

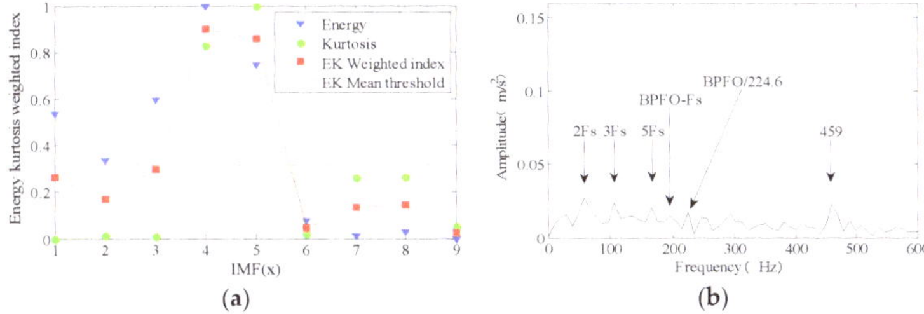

Figure 17. The result of energy kurtosis means filtering. (**a**) Energy kurtosis weighted index of the NASA outer ring IMF components. (**b**) Envelope demodulation of NASA outer ring reconstruction signal.

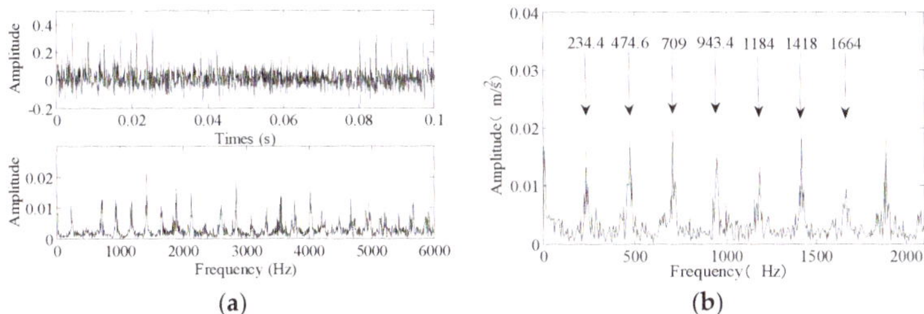

Figure 18. MOMEDA filtered output of NASA outer ring signal and its envelope demodulation. (**a**) Time and frequency domain analysis of MOMEDA filtered output for outer ring signal. (**b**) Envelope demodulation of MOMEDA filtered output for outer ring signal.

MOMEDA effectively extracts the periodic impacts in the reconstructed signal and weakens the frequency components unrelated to faults. It is noteworthy that peaks appear at 234.4 Hz, 474.6 Hz, 709 Hz and 943.4 Hz, etc. These peaks correspond to the outer ring fault frequency 236.4 Hz and its double frequency 472.8 Hz, triple frequency 709.2 Hz, quadrupled frequency 945.6 Hz and so on. After adding MOMEDA, the absolute error between the measured value and theoretical value is 236.4 Hz − 234.4 Hz = 2 Hz, and the relative error is 0.0085, which greatly reduces the frequency error and improves the fault detection accuracy.

4.3. Rolling Element Fault Feature Extraction Application

The time and frequency domain waveforms of the rolling element fault signal are shown in Figure 19a. Compared with the vibration signals of the inner ring and the outer ring, the periodic impacts of the rolling element signal are not obvious. Moreover, the fault frequency is almost submerged by complex frequencies, and it is difficult to extract fault frequency from the frequency domain. Therefore, the rolling element signal is decomposed by FCEEMD, and the effective IMFs are selected by energy kurtosis mean filtering. The results are shown in Figures 19b and 20a.

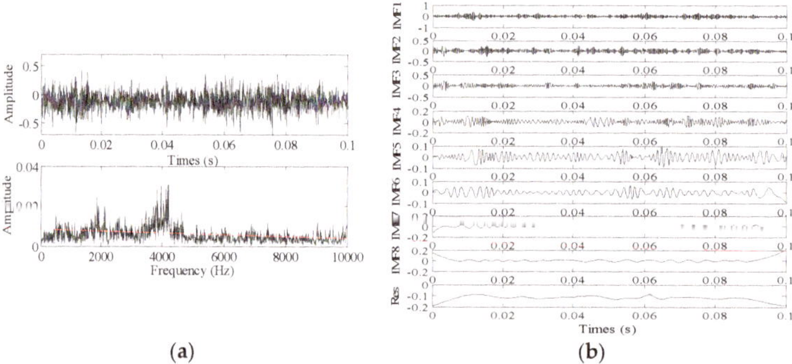

Figure 19. NASA rolling element fault signal and its decomposition results. (**a**) Time and frequency domain analysis of rolling element signal. (**b**) FCEEMD decomposition result of rolling element signal.

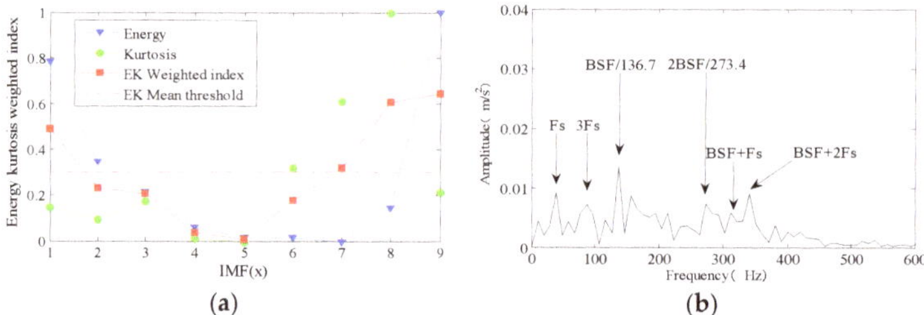

Figure 20. The result of energy kurtosis mean filtering. (**a**) Energy kurtosis weighted index of the NASA rolling element IMFs. (**b**) Envelope demodulation of NASA rolling element reconstruction signal.

The changes in the energy and kurtosis of the IMFs are not regulated, and the energy and kurtosis of the IMFs in the high-frequency band and the low-frequency band are relatively large. According to the energy kurtosis mean filtering, the effective IMFs include IMF1, IMF7, IMF8 and IMF9, and the envelope demodulation of the effective IMFs reconstruction signal is shown in Figure 20b.

Two obvious spectrum peaks appear at 136.7 Hz and 273.4 Hz, which correspond to the rolling element fault frequency 139.9 Hz and its double frequency 279.8 Hz. Due to the influence of noises, the absolute error between the measured value and theoretical value is 139.9 Hz − 136.7 Hz = 3.2 Hz, and the relative error is 0.0229. Furthermore, there are several peaks in the low-frequency part of the envelope spectrum, which correspond to the rotating frequency (Fs = 33.33 Hz) and its frequency multiplication. However, there are not only the modulation signals of fault frequency and rotating frequency in the envelope spectrum but also many interference frequencies, which can cause difficulty in extracting the fault feature. Therefore, we extracted the periodic impacts from the reconstructed signal through MOMEDA. The results are shown in Figure 21a,b.

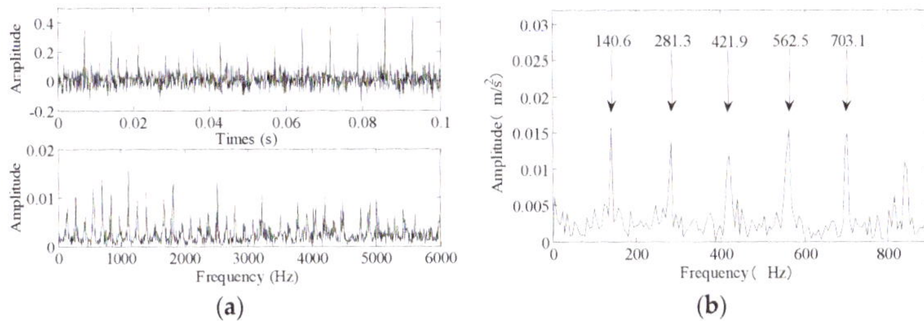

Figure 21. MOMEDA filtered output of NASA rolling element signal and its envelope demodulation. (**a**) Time and frequency domain analysis of MOMEDA filtered output for rolling element signal. (**b**) Envelope demodulation of MOMEDA filtered output for rolling element signal.

Although the impacts of the rolling element vibration signal are not obvious, MOMEDA can still effectively extract the periodic impacts in the reconstructed signal and weaken the frequencies unrelated to the faults. It is noteworthy that the peaks appear at 140.6 Hz, 281.3 Hz, 421.9 Hz and 562.5 Hz, etc. These peaks correspond to the rolling element fault frequency 139.9 Hz and its double frequency 279.8 Hz, triple frequency 419.7 Hz, quadrupled frequency 559.6 Hz and so on. After adding MOMEDA, the absolute error between the measured value and theoretical value is 140.6 Hz − 139.9 Hz = 0.7 Hz, and the relative error is 0.005, which greatly reduces the frequency error. This means the rolling element fault can be diagnosed accurately.

5. Comparison Analysis of Different Feature Extraction Methods

By analyzing and summarizing the results above, we can draw a preliminary conclusion that the fault feature extraction of the rolling element is more difficult compared with the inner ring and outer ring. According to the theory of ref. [40], the frequency spectrum contains the second harmonic of the spin frequency when the rolling element fails. The spin frequency of the rolling element is generated by impacting the inner ring or the outer ring through the rolling element. In general, the rolling element rotates once and produces two impacts. Therefore, the fault frequency of the rolling element is easily submerged by interference frequencies. Most existing methods can achieve better results in feature extraction of the inner ring and outer ring, but they have a poor effect on the feature extraction of the rolling element. These facts prove that the feature extraction of the rolling element is difficult, but the proposed method can achieve good results in the feature extraction of the rolling element.

To verify the superiority of the proposed method, a variety of methods were compared through the rolling element data of the CWRU dataset. In the comparative experiments, the data length was 2048.

5.1. Comparative Analysis of Feature Extraction Methods Based on Different Decomposition Methods

To verify the effectiveness of the MOMEDA demodulation method based on the decomposition method, we compared the method without signal decomposition with the methods based on several signal decomposition strategies. As shown in "MOMEDA" in Figure 22, the envelope spectrum obtained by MOMEDA without signal decomposition contained a large amount of noise. The rolling element envelope spectrum obtained alone by DOMEDA contained many burrs and was coarse, so signal decomposition and noise reduction were indispensable and important processing links. Therefore, we decomposed the rolling element signals by EEMD, FEEMD and FCEEMD. In EEMD, the ratio of the standard deviation of the added noise and original signal (Nstd) was 0.2, and the ensemble member was 10. In FEEMD and FCEEMD, Nstd was 0.2, the number of IMF was 9, the number of screening iterations maxSift was 10 and the rest of the parameters were default values. For the IMFs obtained by each decomposition algorithm, the reconstruction error, root mean square error (RMSE), time-consuming (Time) and standard deviation (SD) are shown in Figure 22a and Table 5. The reconstruction error refers to the deviation between the original signal and the reconstructed signal of all IMFs.

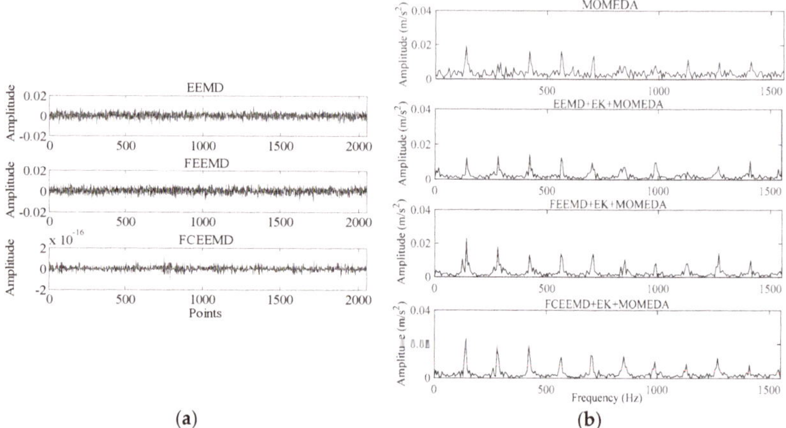

(a) (b)

Figure 22. Comparative analysis of feature extraction methods based on different decomposition methods. (**a**) Reconstruction errors of different decomposition methods. (**b**) Envelope spectrums obtained by different decomposition methods.

Table 5. The performance indexes of decomposition methods.

Method	RMSE	Time (s)	SD
EEMD	0.0027	7.806370	0.1336
FEEMD	0.0027	0.016744	0.1338
FCEEMD	1.8445×10^{-17}	0.013481	0.1337

As shown in Figure 22a, both the reconstruction errors of the EEMD and FEEMD are large, while the reconstruction error (10^{-16}) of the FCEEMD can be neglected. The results show that the FCEEMD can decompose the signal into several IMFs completely, there is no energy leakage and RMSE also demonstrates this advantage of FCEEMD. In terms of being time-consuming, the decomposition efficiency of FEEMD and FCEEMD was high compared with EEMD, so they were suitable for real-time data processing. From the standard deviation of the reconstructed signal, the decomposed signals were stable and reliable. Therefore, FCEEMD had the characteristics of rapidity and high accuracy.

Similarly, we selected the effective IMFs by energy kurtosis mean filtering and extracted the periodic impacts in the effective IMFs reconstructed signal through the MOMEDA. The envelope demodulations of the periodic impacts are shown in Figure 22b. All three methods obtained the fault frequencies of the rolling element and its harmonics. It is noteworthy that the amplitude of the envelope obtained by the EEMD method was small, and the spectrum peaks of the high-frequency band were hardly recognized. The spectrum peaks obtained by FEEMD and FCEEMD were relatively obvious, and the spectrum peaks of the high-frequency band were easily identified. In addition, the envelope spectrum obtained by FCEEMD was cleaner than that obtained by FEEMD, and the interference frequencies were less. Therefore, the feature extraction method based on FCEEMD was more efficient than others.

5.2. Comparative Analysis of Feature Extraction Methods Based on Different Effective IMFs Selection Methods

In the second experiment, the proposed method was compared with the method in ref. [36]. Firstly, the rolling element signal was decomposed by EEMD, and then the

IMFs that had a large correlation with the original signal were selected as effective IMFs, including IMF1, IMF2, IMF3, IMF4, IMF5 and IMF6. The results are shown in Figure 23a and Table 6.

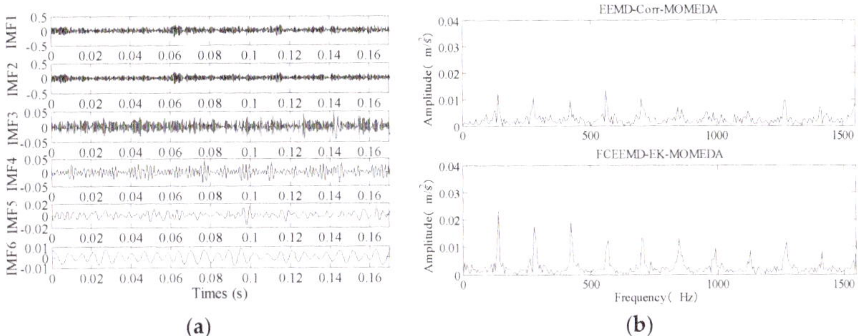

Figure 23. EEMD decomposition results of the rolling element fault signal and envelope spectrum of the two methods. (**a**) EEMD decomposition results of rolling element fault sign. (**b**) Envelope spectrum of the two methods.

Table 6. Correlation coefficient of IMF components.

IMF1	IMF2	IMF3	IMF4	IMF5	IMF6	IMF7	IMF8
0.999	0.948	0.223	0.233	0.111	0.077	0.010	0.005

Then, the impacts contained in the reconstruction signal of IMF1~IMF6 were extracted by MOMEDA. The envelope spectrum of the impacts was compared with that obtained by the proposed method, as shown in Figure 23b.

Compared with the proposed method, the results of the method based on EEMD, and correlation coefficient were not good, the envelope spectrum was affected by many interference frequencies and the characteristic frequencies of high-frequency bands were almost submerged by irrelevant components. The effective IMFs were selected by the correlation coefficient in the method, and those IMFs which had a strong correlation with the original signal were included in the reconstructed signal. Therefore, interference noises and fault signals may be included in the reconstructed signal. In addition, EEMD is time-consuming, and the selection of the correlation threshold is based on experience, which leads to a lack of adaptability. By introducing the energy kurtosis weighted index, the proposed method considered the signal strength and fault impact characteristics of the IMF synthetically. Meanwhile, the mean of the energy kurtosis weighted indices of all IMFs was used as the threshold. The determination of the threshold was adaptive, and the results show that energy kurtosis mean filtering was more effective than the correlation coefficient method.

5.3. Comparative Analysis of Feature Extraction Methods Based on Different Deconvolution Methods

In the third experiment, the influence of different deconvolution methods on the results was studied. Firstly, the rolling element signal was decomposed by FCEEMD. Then, the effective IMFs were selected by energy kurtosis mean filtering. Finally, the impacts contained in the reconstructed signals were extracted by MED, MCKD and MOMEDA, respectively. The filter length and iteration number of MED and MCKD were 500 and 30, respectively, and the period of the MCKD was Tb = fs/fb = 12,000/141.1693 = 85.004 [21]; the number of iterations was 30. The time domain diagram and envelope spectrum of the impact signals obtained by the three methods are shown in Figure 24a,b.

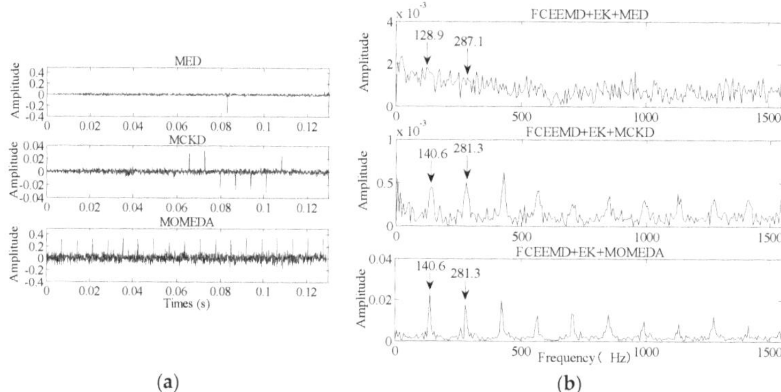

(a) (b)

Figure 24. The time domain diagram and envelope spectrum of the impact signals obtained by the three methods. (**a**) Time domain diagram of three groups of impact signals. (**b**) Envelope demodulation of three groups of impact signals.

The impact signal extracted by MED contained only one impact, which led to the loss of the periodic impacts. Therefore, the fault frequency of the rolling element was submerged in the interference frequencies, and the fault could not be identified. The impact signal extracted by MCKD contained a series of impacts, but these impacts did not have continuous periodicity. Moreover, the envelope spectrum obtained by MCKD contained a lot of interference frequencies, which affected the results. The impact signal extracted by MOMEDA was one of the periodic impacts, and the envelope spectrum contained many clear spectrum peaks. Therefore, MOMEDA can extract the periodic impacts of the reconstructed signal more effectively compared with MED and MCKD. Through the analysis above, the superiority of the proposed method was verified.

5.4. Noise Robustness Analysis of Feature Extraction Method

In the fourth experiment, we analyzed the feature extraction performance of the method under different noise levels. First, four levels of Gaussian white noise with signal-to-noise ratio (SNR) of 10 dB, 15 dB, 20 dB, and 25 dB were added to the rolling element vibration signal, respectively, and four rolling element signals with different degrees of noise pollution were obtained. The time-domain and frequency-domain waveforms of four rolling element signals with different noise levels are shown in Figure 25. Many irregular points and burrs were added to both the time and frequency domain waveforms, and the difficulty of extracting fault feature frequencies was also increased.

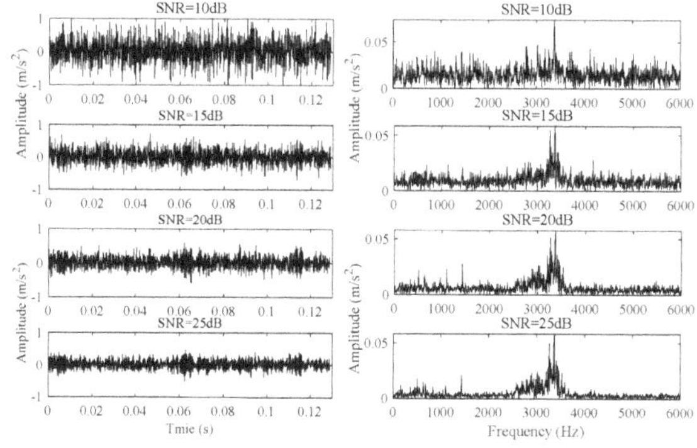

Figure 25. The time-domain and frequency-domain waveforms of four rolling element signals with different noise levels.

In order to verify the noise robustness of the proposed method, we analyzed the rolling element vibration signal under different levels of noise by the proposed method. The feature extraction results are shown in Figure 26. Overall, the envelope curves obtained by feature extraction at four different noise levels were very clear, with very few noise points and burrs. In addition, the rolling element faults characteristic frequencies, and their high multipliers were very obvious under four different noise levels, and the types of rolling element faults were accurately identified. Therefore, under different noise levels, the proposed fault feature extraction method can effectively extract the rolling element fault feature frequency. Our experimental results show that the proposed feature extraction method has good noise robustness.

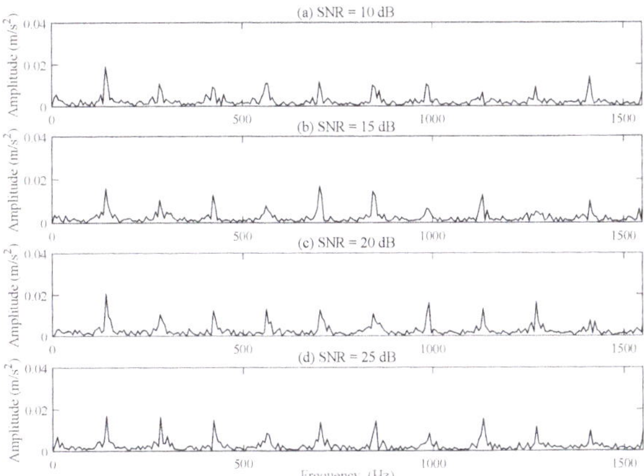

Figure 26. The envelope spectrum obtained by the proposed method under different noise levels. (**a**) Envelope spectrum at noise level SNR = 10 dB. (**b**) Envelope spectrum at noise level SNR = 15 dB. (**c**) Envelope spectrum at noise level SNR = 20 dB. (**d**) Envelope spectrum at noise level SNR = 25 dB.

5.5. Comparison of the Obtained Results with Literature

Similar to the fault feature extraction results listed in reference [41], it is particularly important to compare the results obtained from different literature. Table 7 shows the studies on bearing fault feature extraction. Very few metrics can be used to evaluate the quality of the envelope demodulation effect, so we qualitatively classified the envelope demodulation effect into "very obvious", "obvious" and "general" according to the envelope waveform. In all the literature, the method we proposed achieved good results, and the envelope demodulation result was defined as "very obvious". As can be seen from Table 7, the proposed method was more successful than methods in other pieces of literature.

Table 7. The reported studies on the bearing fault.

Author	Model	Dataset	Fault Frequency Waveform
Yin [42]	Improved ensemble noise reconstructed EMD	XJTU-SY, IMS	obvious
Li [43]	Improved EEMD based on improved adaptive resonance technology	CWRU, Experimental of authors	obvious, general
Han [44]	Teager energy operator and CEEMD	Wind turbine experimental test rig	general
Zhou [25]	Adaptive VMD and MOMEDA	CWRU, IMS	very obvious
Xiao [26]	Adaptive MOMEDA and Teager energy operator	Experimental of authors	very obvious
Wang [45]	Resonance sparse decomposition and improved MOMEDA	Experimental of authors	very obvious
Li [24]	Hankel matrix, signal spectrum is reconstructed, MOMEDA	CWRU	obvious
Authors of this article	FCEEMD Energy Kurtosis Mean Filtering, MOMEDA	CWRU, IMS	very obvious

6. Conclusions

Bearings are easily damaged in harsh industrial environments, so the study of bearing fault feature extraction methods is particularly important. To solve the problems of low efficiency and residual noise in EEMD and FEEMD and extract the fault frequencies on the basis of obtaining effective IMFs, a fast, complementary ensemble empirical mode decomposition (FCEEMD) energy kurtosis mean filtering-based fault feature extraction method was proposed. By analyzing the CWRU and NASA bearing datasets, the validity and superiority of the proposed method were proved, and the following conclusions can be drawn.

(1)   In most signal noise reduction processes, EMD, EEMD and VMD are the most commonly used decomposition methods, which are not only too slow in the decomposition process but also make it difficult to select parameters. Compared with EEMD and FEEMD, the proposed FCEEMD had the smallest reconstruction error, the lowest time consumption and the most stable intrinsic mode functions (IMF), which indicates that the introduced pairs of white noise with opposite signs can completely neutralize the residual white noise in FEEMD and suppress the mode aliasing between IMF.

(2)   Among most of the existing IMF selection methods, the correlation coefficient or kurtosis is the most commonly used method. Compared with the IMF selection methods based on a single index such as correlation coefficient, energy or kurtosis, the reconstructed signal obtained by the proposed method contained more fault-related shock signals and less noise, which shows that the proposed energy kurtosis mean filtering can comprehensively consider the fault impact and signal strength.

(3)   In most of the deconvolution methods based on minimum entropy deconvolution (MED) and maximum correlation kurtosis deconvolution (MCKD), the decomposition module and the effective IMF selection module are not added to the method. Although these methods have achieved good results in the feature extraction of the inner and outer rings, the rolling element fault feature extraction effect is not very good. Compared with the feature extraction methods based on MED and MCKD, the proposed MOMEDA-based method can obtain periodic impact signals of rolling elements, and the fault feature frequency is more obvious and precise, which shows that the combined method based on MOMEDA can extract complex fault frequency characteristics more effectively.

The proposed method has good efficiency and reliability, and it is expected to be applied to the feature extraction of rotating machinery in actual production. It is worth noting that the proposed method has a relatively long process, which is not conducive to improving the real-time reliability of fault detection. In future production, the simplification of the process and the improvement of reliability will be new research points.

Author Contributions: Conceptualization, C.Z. and L.X.; methodology, C.Z.; software, L.X.; validation, Y.J., S.W. and Z.Z.; formal analysis, C.Z.; investigation, C.Z.; resources, L.X.; data curation, C.Z.; writing—original draft preparation, Y.J.; writing—review and editing, C.Z.; visualization, C.Z.; supervision, Y.J. and S.W.; project administration, L.X.; funding acquisition, L.X. All authors have read and agreed to the published version of the manuscript.

Funding: This work is supported by the National Natural Science Foundation of China (No. 51765022 and No. 41971392), the Ph.D. research startup foundation of Yunnan Normal University (No. 01000205020503131), the Fundamental Research Program of Yunnan Province (No. 202201AU070055) and the Project of Educational Commission of Yunnan Province of China.

Institutional Review Board Statement: Not applicable.

Informed Consent Statement: Informed consent was obtained from all subjects involved in the study.

Data Availability Statement: The data used to support the findings of this study are available from the corresponding author upon request.

Acknowledgments: The authors sincerely thank the team for their guidance and thank the Case West Reserve University and the University of Cincinnati for their bearing datasets. The authors also sincerely thank the reviewers for their comments and suggestions, which contributed to the improvement of the paper.

Conflicts of Interest: The authors declare no conflict of interest.

References

1. Cheng, J.; Yang, Y.; Li, X.; Cheng, J. Adaptive periodic mode decomposition and its application in rolling bearing fault diagnosis. *Mech. Syst. Signal Process.* **2021**, *161*, 107943. [CrossRef]
2. Chen, Z.; Guo, L.; Gao, H.; Yu, Y.; Wu, W.; You, Z.; Dong, X. A fault pulse extraction and feature enhancement method for bearing fault diagnosis. *Measurement* **2021**, *182*, 109718. [CrossRef]
3. Ge, B.; Wang, Z.-C.; Ding, Y.-J.; Mo, Y. Hilbert square demodulation and error mitigation of the measured nonlinear structural dynamic response. *Mech. Syst. Signal Process.* **2021**, *160*, 107935. [CrossRef]
4. Liu, Z.; Peng, D.; Zuo, M.J.; Xia, J.; Qin, Y. Improved Hilbert–Huang transform with soft sifting stopping criterion and its application to fault diagnosis of wheelset bearings. *ISA Trans.* **2021**, *125*, 426–444. [CrossRef]
5. Laval, X.; Mailhes, C.; Martin, N.; Bellemain, P.; Pachaud, C. Amplitude and phase interaction in Hilbert demodulation of vibration signals: Natural gear wear modeling and time tracking for condition monitoring. *Mech. Syst. Signal Process.* **2021**, *150*, 107321. [CrossRef]
6. Huang, N.; Shen, Z.; Long, S.; Wu, M.L.C.; Shih, H.; Zheng, Q.; Yen, N.-C.; Tung, C.-C.; Liu, H. The empirical mode decomposition and the Hilbert spectrum for nonlinear and non-stationary time series analysis. *Proc. R. Soc. Lond. Ser. A Math. Phys. Eng. Sci.* **1998**, *454*, 903–995. [CrossRef]
7. Dominique, K.; Zosso, D. Variational Mode Decomposition. *IEEE Trans. Signal Processing A Publ. IEEE Signal Processing Soc.* **2014**, *62*, 531–544.
8. Wu, Z.; Huang, N. Ensemble empirical mode decomposition: A noise-assisted data analysis method. *Adv. Adapt. Data Anal.* **2009**, *1*, 1–41. [CrossRef]
9. Yeh, J.R.; Shieh, J.S.; Huang, N.E. Complementary ensemble empirical mode decomposition: A novel noise enhanced data analysis method. *Adv. Adapt. Data Anal.* **2010**, *2*, 135–156. [CrossRef]
10. Gao, K.; Xu, X.; Li, J.; Jiao, S.; Shi, N. Weak fault feature extraction for polycrystalline diamond compact bit based on ensemble empirical mode decomposition and adaptive stochastic resonance. *Measurement* **2021**, *178*, 109304. [CrossRef]
11. Gu, J.; Peng, Y. An improved complementary ensemble empirical mode decomposition method and its application in rolling bearing fault diagnosis. *Digit. Signal Process.* **2021**, *113*, 103050. [CrossRef]
12. Li, Y.-X.; Wang, L. A novel noise reduction technique for underwater acoustic signals based on complete ensemble empirical mode decomposition with adaptive noise, minimum mean square variance criterion and least mean square adaptive filter. *Def. Technol.* **2020**, *16*, 543–554. [CrossRef]
13. Li, Y.; Wang, L.; Li, X.; Yang, X. A novel linear spectrum frequency feature extraction technique for warship radio noise based on complete ensemble empirical mode decomposition with adaptive noise, duffing chaotic oscillator, and weighted-permutation entropy. *Entropy* **2019**, *21*, 507. [CrossRef] [PubMed]
14. Zhang, J.; Feng, F.; Marti-Puig, P.; Caiafa, C.F.; Sun, Z.; Duan, F.; Solé-Casals, J. Serial-EMD: Fast empirical mode decomposition method for multi-dimensional signals based on serialization. *Inf. Sci.* **2021**, *581*, 215–232. [CrossRef]
15. Wang, Y.-H.; Yeh, C.-H.; Young, H.-W.V.; Hu, K.; Lo, M.-T. On the computational complexity of the empirical mode decomposition algorithm. *Phys. A Stat. Mech. Its Appl.* **2014**, *400*, 159–167. [CrossRef]
16. Chegini, S.N.; Manjili, M.J.H.; Bagheri, A. New fault diagnosis approaches for detecting the bearing slight degradation. *Meccanica* **2020**, *55*, 261–286. [CrossRef]
17. Sun, N.; Zhou, J.; Liu, G.; He, Z. A hybrid wind speed forecasting model based on a decomposition method and an improved regularized extreme learning machine. *Energy Procedia* **2019**, *158*, 217–222. [CrossRef]
18. Xia, J.; Zhao, L.; Bai, Y.; Yu, M.; Wang, Z. Feature extraction for rolling element bearing weak fault based on MCKD and VMD. *J. Vib. Shock.* **2017**, *36*, 78–83.
19. Duan, R.; Liao, Y.; Yang, L.; Xue, J.; Tang, M. Minimum entropy morphological deconvolution and its application in bearing fault diagnosis. *Measurement* **2021**, *182*, 109649. [CrossRef]
20. He, Z.; Chen, G.; Hao, T.; Liu, X.; Teng, C. An optimal filter length selection method for MED based on autocorrelation energy and genetic algorithms. *ISA Trans.* **2021**, *109*, 269–287. [CrossRef]
21. Zhang, Y.; Lv, Y.; Ge, M. Time–frequency analysis via complementary ensemble adaptive local iterative filtering and enhanced maximum correlation kurtosis deconvolution for wind turbine fault diagnosis. *Energy Rep.* **2021**, *7*, 2418–2435. [CrossRef]
22. Cui, H.; Guan, Y.; Chen, H. Rolling element fault diagnosis based on VMD and sensitivity MCKD. *IEEE Access* **2021**, *9*, 120297–120308. [CrossRef]
23. McDonald, G.L.; Zhao, Q. Multipoint optimal minimum entropy deconvolution and convolution fix: Application to vibration fault detection. *Mech. Syst. Signal Process.* **2017**, *82*, 461–477. [CrossRef]

24. Li, Y.; Cheng, G.; Liu, C. Research on bearing fault diagnosis based on spectrum characteristics under strong noise interference. *Measurement* **2021**, *169*, 108509. [CrossRef]
25. Zhou, X.; Li, Y.; Jiang, L.; Zhou, L. Fault feature extraction for rolling bearings based on parameter-adaptive variational mode decomposition and multi-point optimal minimum entropy deconvolution. *Measurement* **2021**, *173*, 108469. [CrossRef]
26. Xiao, C.; Tang, H.; Ren, Y.; Xiang, J.; Kumar, A. Adaptive MOMEDA based on improved advance-retreat algorithm for fault features extraction of axial piston pump. *ISA Trans.* **2022**, *128*, 503–520. [CrossRef] [PubMed]
27. Bayram, S.; Kaplan, K.; Kuncan, M.; Ertunc, H.M. The effect of bearings faults to coefficients obtaned by using wavelet transform. In Proceedings of the 2014 22nd Signal Processing & Communications Applications Conference, Trabzon, Turkey, 23–25 April 2014.
28. Kaya, Y.; Kuncan, M.; Kaplan, K.; Minaz, M.R.; Ertunç, H. A new feature extraction approach based on one dimensional gray level co-occurrence matrices for bearing fault classification. *J. Exp. Theor. Artif. Intell.* **2021**, *33*, 1735530. [CrossRef]
29. Liu, Q.; Yang, J.; Zhang, K. An improved empirical wavelet transform and sensitive components selecting method for bearing fault. *Measurement* **2022**, *187*, 110348. [CrossRef]
30. Kaya, Y.; Kuncan, F.; ErtunÇ, H.M. A new automatic bearing fault size diagnosis using time-frequency images of CWT and deep transfer learning methods. *Turk. J. Electr. Eng. Comput. Sci.* **2022**, *30*, 1851–1867. [CrossRef]
31. Kuncan, M. An Intelligent Approach for Bearing Fault Diagnosis: Combination of 1D-LBP and GRA. *IEEE Access* **2020**, *8*, 137517–137529. [CrossRef]
32. Han, Y.; Li, K.; Ge, F.; Wang, Y.A.; Xu, W. Online fault diagnosis for sucker rod pumping well by optimized density peak clustering. *ISA Trans.* **2022**, *120*, 222–231. [CrossRef] [PubMed]
33. Zheng, J.; Tong, J.; Ni, Q.; Pan, H. Partial ensemble approach to resolve the mode mixing of extreme-point weighted mode decomposition. *Digit. Signal Process.* **2019**, *89*, 70–81. [CrossRef]
34. Lin, B.; Zhang, C. A novel hybrid machine learning model for short-term wind speed prediction in inner Mongolia, China. *Renew. Energy* **2021**, *179*, 1565–1577. [CrossRef]
35. Xue, X.; Zhou, J. A hybrid fault diagnosis approach based on mixed-domain state features for rotating machinery. *ISA Trans.* **2017**, *66*, 284–295. [CrossRef]
36. Wang, Z.; Wang, J.; Zhang, J.; Zhao, Z.; Kou, Y. Fault diagnosis of gearbox based on improved MOMEDA. *J. Vib. Meas. Diagn.* **2018**, *38*, 176–181.
37. Guo, Y.; Zhao, X.; Shangguan, W.-B.; Li, W.; Lü, H.; Zhang, C. Fault characteristic frequency analysis of elliptically shaped bearing. *Measurement* **2020**, *155*, 107544. [CrossRef]
38. Li, Z.; Ma, J.; Wang, X.; Li, X. An Optimal Parameter Selection Method for MOMEDA Based on EHNR and Its Spectral Entropy. *Sensors* **2021**, *21*, 533. [CrossRef]
39. Yang, C.; Ma, J.; Wang, X.; Li, X.; Li, Z.; Luo, T. A novel based-performance degradation indicator RUL prediction model and its application in rolling bearing. *ISA Trans.* **2021**, *121*, 349–364. [CrossRef]
40. Kulkarni, V.V.; Khadersab, A. Spectrum analysis of rolling element bearing faults. *Mater. Today Proc.* **2021**, *46*, 4679–4684. [CrossRef]
41. Kaplan, K.; Kaya, Y.; Kuncan, M.; Mïnaz, M.R.; Ertunç, H.M. An improved feature extraction method using texture analysis with LBP for bearing fault diagnosis. *Appl. Soft. Comput.* **2020**, *87*, 106019. [CrossRef]
42. Yin, C.; Wang, Y.; Ma, G.; Wang, Y.; Sun, Y.; He, Y. Weak fault feature extraction of rolling bearings based on improved ensemble noise-reconstructed EMD and adaptive threshold denoising. *Mech. Syst. Signal Process.* **2022**, *171*, 108834. [CrossRef]
43. Li, H.; Liu, T.; Wu, X.; Li, S. Research on test bench bearing fault diagnosis of improved EEMD based on improved adaptive resonance technology. *Measurement* **2021**, *185*, 109986. [CrossRef]
44. Han, T.; Liu, Q.; Zhang, L.; Tan, A.C.C. Fault feature extraction of low speed roller bearing based on Teager energy operator and CEEMD. *Measurement* **2019**, *138*, 400–408. [CrossRef]
45. Wang, C.; Li, H.; Ou, J.; Hu, R.; Hu, S.; Liu, A. Identification of planetary gearbox weak compound fault based on parallel dual-parameter optimized resonance sparse decomposition and improved MOMEDA. *Measurement* **2020**, *165*, 108079. [CrossRef]

Article

Fault Diagnosis of Check Valve Based on KPLS Optimal Feature Selection and Kernel Extreme Learning Machine

Xuyi Yuan [1], Yugang Fan [1,*], Chengjiang Zhou [2,*], Xiaodong Wang [1] and Guanghui Zhang [1]

[1] Faculty of Information Engineering and Automation, Kunming University of Science and Technology, Kunming 650500, China
[2] School of Information Science and Technology, Yunnan Normal University, Kunming 650500, China
* Correspondence: ygfan@kust.edu.cn (Y.F.); chengjiangzhou@foxmail.com (C.Z.)

Abstract: The check valve is the core part of high-pressure diaphragm pumps. It has complex operation conditions and has difficulty characterizing fault states completely with its single feature. Therefore, a fault signal diagnosis model based on the kernel extreme learning machine (KELM) was constructed to diagnose the check valve. The model adopts a multi-feature extraction method and reduces dimensionality through kernel partial least squares (KPLS). Firstly, we divided the check valve vibration signal into several non-overlapping samples. Then, we extracted 16 time-domain features, 13 frequency-domain features, 16 wavelet packet energy features, and energy entropy features from each sample to construct a multi-feature set characterizing the operation state of the check valve. Next, we used the KPLS method to optimize the 45 dimension multi-feature data and employed the processed feature set to establish a KELM fault diagnosis model. Experiments showed that the method based on KPLS optimal feature selection could fully characterize the operating state of the equipment with an accuracy rate of 96.88%. This result indicates the high accuracy and effectiveness of the multi-feature set constructed with the KELM fault diagnosis model.

Keywords: KPLS; KELM; fault diagnosis; check valve

Citation: Yuan, X.; Fan, Y.; Zhou, C.; Wang, X.; Zhang, G. Fault Diagnosis of Check Valve Based on KPLS Optimal Feature Selection and Kernel Extreme Learning Machine. *Coatings* **2022**, *12*, 1320. https://doi.org/10.3390/coatings12091320

Academic Editors: Ke Feng and Luis Norberto López De Lacalle

Received: 23 July 2022
Accepted: 8 September 2022
Published: 10 September 2022

Publisher's Note: MDPI stays neutral with regard to jurisdictional claims in published maps and institutional affiliations.

1. Introduction

As the core piece of equipment in ore transportation pipelines, the check valve directly affects the operation of pipeline systems through its operation status. Research on the fault diagnosis method for check valves is of great significance for the development of the pipeline transportation industry. The operating conditions of check valves are complex, with the vibration signal being a periodic pulse signal affected by environmental noise and other factors. When failure occurs, signal characteristics experience interference and are challenging to extract. Thus, a single feature of a check valve cannot fully characterize the operating state of the equipment.

Characterizing operating states extracting signal features is the basis for fault diagnosis of mechanical equipment. Chen et al. [1] used a continuous wavelet transform (CWT) to preprocess an original vibration signal and constructed a fused convolutional neural network (CNN) with a square pool structure to extract signal features and to realize fault diagnosis of mechanical equipment. Peng et al. [2] proposed a fault classification method based on multi-feature extraction and an improved Mahalanobis–Taguchi System (MTS). The method involves extracting multi-dimensional features using complete ensemble empirical mode decomposition with adaptive noise (CEEMDAN) for time, frequency, and adaptive white noise. The authors constructed a DAG-MTS multi-classification model based on the characteristics of the MTS system and a directed acyclic graph (DAG) and applied it to a bearing's fault diagnosis. To solve the problem of noise in a diesel engine's vibration signal and address the difficulty of feature extraction, Jiang et al. [3] proposed a diesel engine fault diagnosis method focusing on the extraction of the wavelet packet energy spectrum and the selection of fuzzy entropy features. The fuzzy entropy selects components out of the feature set extracted from the wavelet packet energy spectrum and inputs the selected features into the least squares twin support vector machine (LSSVM) for fault diagnosis.

Constructing a feature set that characterizes the operating state of equipment and using it to establish a fault diagnosis model is the key to the fault diagnosis of mechanical equipment. Ding et al. [4] proposed a method for scintillation detector fault diagnosis based on the extreme learning machine (ELM), and this method could not only classify the faults of the failed detector but also intelligently determine the severity of various faults. Lee et al. [5] proposed a novel remaining useful life (RUL) estimation method based on systematic feature engineering and the extreme learning machine (ELM) for seven out of eleven bearings; the proposed method reduced the mean absolute error (MAE), root mean square error (RMSE), and mean absolute percentage error (MAPE) in the RUL estimation by over 50%. Pang et al. [6] proposed an ensembled kernel extreme learning machine that fuses multi-domain features. It selects the most suitable stacked noise reduction autoencoder method through time-domain and frequency-domain feature extraction. It uses a kernel extreme learning machine with deep features for rotating machinery fault diagnosis. Wang et al. [7] proposed an ensemble extreme learning machine (EELM) that consists of two heterogeneous ELM networks. First, it displays the target data using a clustering algorithm. Then, it applies Gaussian-style activation between each target as an input to the back-end classifier to propose a non-empirically specified threshold based on the EELM multi-label classifier. Later, the multiple binary classifiers are combined for composite fault diagnosis. Zhang et al. [8] introduced an online fault diagnosis method that changes the fixed structure of the extreme learning machine into an elastic structure using incremental support vector data description (ISVDD) and an extreme learning machine with an incremental output structure (IOELM). The ISVDD is used to detect a new failure mode, while the IOELM is used to recognize the specific failure mode. Harishvijey et al. [9] proposed an automatic signal classification method for detecting seizures from an EEG signal using an empirical wavelet transform (EWT) feature extraction method, K-principal component analysis (K-PCA)-based feature reduction, and a fuzzy logic-embedded RBF kernel-based ELM. Shen et al. [10] proposed a feature selection and fusion method based on the poll mode and optimized WKPCA. Considering the variation in fault information collected by different sensors, the diagnosis rate in the extreme learning machine (ELM) is taken as the index for the evaluation of each single sensor, and then the sensitivity weight matrix of the features extracted by multiple sensors is constructed after linear normalization. Based on the screened temperature-sensitive points and measured thermal displacement data, an optimized extreme learning machine based on the marine predator algorithm (MPA-ELM) was developed to predict the thermal displacement of an electric spindles model [11].

Huang et al. [12] introduced a kernel function into the ELM to replace the simplicity of the hidden layer. They proposed the KELM, which significantly improves the generalization performance of the ELM. Moreover, Huang et al.'s [13] research on the improvement of the ELM further enhanced the stability, sparsity, and accuracy of the algorithm under both general and specific conditions and accelerated the training speed of the ELM. Chen et al. [14] proposed a new fault diagnosis method based on hierarchical machine learning based on the KELM. A grid search strategy with cross-validation was used to optimize the parameters of hierarchical machine learning (HML). It was applied to detect and identify rotating machinery faults, obtaining excellent results. Based on the KELM, Su et al. [15] added self-adaptive particle swarm optimization (SAPSO) to optimize its parameters and proposed a fault diagnosis method for rotary bearings under mixed working conditions.

The fault diagnosis methods above provide a helpful reference for detecting the operating state of a check valve. However, as check valve operating conditions are complex due to the operating conditions of industrial production and environmental noise, the vibration signal is affected, thus resulting in nonlinear and non-stationary signals. Due to these characteristics, it is difficult to extract the fault characteristics of components comprehensively and accurately with only a single time-domain, frequency-domain/or time-frequency-domain method. Therefore, this paper proposes a new, multi-feature fault feature extraction method for check valves to characterize their fault state information comprehensively.

Based on multi-feature sets that characterize the operating state of one-way valves, a fault diagnosis model based on KELM was established. The model first constructs the multi-feature set to determine the fault state of the one-way valve comprehensively and accurately and then uses the multi-feature set to train the KELM model to diagnose the check valve fault conditions.

2. Basic Principles

2.1. Feature Extraction

2.1.1. Time-Domain Features

In fault diagnosis, feature fault parameters are generally sensitive to the different information from various states. Usually, fault feature extraction has no specific restrictions on the number and types of feature parameters, so features with high sensitivity to fault information differences and strong reliability are often selected as fault features [16]. The time-domain characteristics of a signal can directly reflect the dynamic changes in the signal's time domain, and they can characterize the fault types of bearings and check valves. Since there may be one or more characteristic parameters corresponding to different states, the selection of the fault feature parameters follows the principles of high sensitivity, high reliability, and feasibility. In this paper, 16 time-domain feature statistics were used and shown in Table 1 below.

2.1.2. Frequency-Domain Features

To characterize the relationship between the frequency and amplitude of a vibration signal, the signal can be transform into the frequency domain with a Fourier transform and the frequency domain characteristics of the signal analyzed in the frequency domain. Frequency domain analysis methods include amplitude spectrum and power spectrum analyses. The frequency-domain characteristic statistics used in this study are shown in Table 1 [17].

Table 1. Composition of multi-domain feature sets.

Multi-Domain Category	Number	Remark
Time-domain features	16	Peak value, mean value, root mean square value, variance, standard deviation, fourth-order center moment, peak factor, kurtosis, pulse factor, margin, waveform factor, the center of gravity frequency, mean square frequency, frequency variance, root mean square frequency, frequency standard deviation
Frequency-domain features	13	Mean, center of gravity frequency, average frequency, maximum value, average phase angle, energy, power, root variance frequency, root mean square frequency, root variance amplitude, maximum phase angle, phase angle range
Time-frequency-domain features	16	Wavelet packet energy features, wavelet packet energy entropy features
Multi-domain features	45	

2.1.3. Energy Characteristics of Wavelet Packets

As a linear transformation method, wavelet packet transform [18] satisfies the law of conservation of energy [19], which is:

$$\int_{-\infty}^{+\infty} |f(t)|^2 \, dt = \sum_j \sum_k |c_j, k|^2 \tag{1}$$

Since the wavelet packet coefficient contains the dimension of energy, it is used in energy analysis to determine each frequency band's energy size according to the signal's wavelet packet coefficient. Time-frequency-domain analysis methods include the short-time Fourier transform, S transform, and wavelet transform methods.

2.1.4. Energy Entropy Characteristics of Wavelet Packets

Entropy is an index used to measure the degree of disorder of information. The higher the value, the higher the disorder degree of disorder of information is and the smaller the contribution to precision is. When the entropy is smaller, the information contribution is more prominent, and the degree of disorder of information is lower.

If we perform layer j wavelet packet decomposition on the signal, assuming that the decomposition sequence is $X_{i,j}$ and $E_{i,j}$ is the signal sequence energy, then the probability density of the frequency band energy is:

$$P(X_{i,j}) = \frac{E_{i,j}}{E_j} \quad E_j = \sum_{i=1}^{2^j} E_{i,j} \quad \sum_{i=1}^{2^j} P(X_{i,j}) = 1 \tag{2}$$

Then, the band energy entropy of the wavelet packet decomposition [20] is:

$$W_{EE} = -\sum_{i=1}^{n} p(X_{i,j}) \log_2(X_{i,j}) \tag{3}$$

Through analysis, it can be found that the larger the band energy entropy of the wavelet packet decomposition is, the more random the distribution of energy in each band is. The lower the number of bands containing energy and the smaller the W_{EE}, the more regular the energy distribution is.

The multi-domain feature adopted in this paper contained 45 dimensions for the feature components.

2.2. Core Extreme Learning Machine—KELM

The standard extreme learning machine [21] (ELM) is composed of three layers: the input layer, hidden layer, and output layer, respectively. It functions based on single hidden layer feedforward neural networks (SLFNs), but the hidden layer of the SLFN only has a one-layer backpropagation (BP) neural network. The topological structure of the ELM is shown in Figure 1.

As an efficient single hidden layer feedforward neural network, we assume that, for a given n training samples $X = (x_1, x_2, \ldots, x_n) \in R^{d1 \times n}$, the labels are $Y = (y_1, y_2, \ldots, y_n) \in R^{n \times d2}$, where d_1 and d_2 represent the dimensions of the input data and output data, respectively. The weight $W = \omega_{ij} \in R^{d1 \times L}$ of the hidden layer of the ELM is randomly selected, where L represents the number of neurons in the hidden layer. The calculation of the hidden layer is the same as the calculation of traditional forward propagation networks with $H = g(X, W)$, where $H \in R^{n \times L}$ and $g(\cdot)$ are the activation functions.

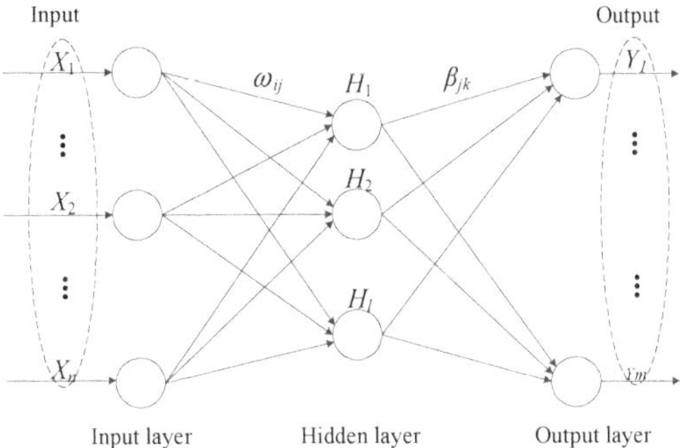

Figure 1. Extreme learning machine model.

The learning objective of an extreme learning machine is to solve the output weight β by minimizing the sum of prediction error loss functions. The objective function is:

$$\min L_{ELM} = \frac{1}{2}\|\beta\|^2 + \frac{C}{2}\|Y - H\beta\|^2 \tag{4}$$

where the C value directly affects the generalization performance of ELM and is a regularization coefficient.

$$H = \begin{bmatrix} g(w_1,b_1,x_1) & \cdots & g(w_L,b_L,x_1) \\ \vdots & & \vdots \\ g(w_1,b_1,x_N) & \cdots & g(w_L,b_L,x_N) \end{bmatrix}_{N\times L} \tag{5}$$

$$\beta = \left[\beta_1^T, \beta_2^T \dots, \beta_L^T\right]_{L\times m}^T \tag{6}$$

$$Y = \left[y_1^T, y_2^T \dots, y_L^T\right]_{N\times m}^T \tag{7}$$

Take the derivative β of Equation (1) and set it to 0, and the calculation formula of the output weight β can be obtained as follows:

$$\beta = \begin{cases} (\frac{I}{C}+H^TH)^{-1}H^TY, & N\geq L \\ H^T(\frac{I}{C}+HH^T)^{-1}Y, & N<L \end{cases} \tag{8}$$

where I is the identity matrix.

Compared with the ELM, the KELM introduces the kernel function [12], which replaces the feature mapping of the hidden layer in ELM. The idea is to map the input sample data to the high-dimensional space and replace the inner product operation in the transformed high-dimensional space with the kernel operation in the original input space [22].

The composition of the kernel matrix Ω_{ELM} is as follows:

$$\begin{cases} \Omega_{ELM}=HH^T \\ \Omega_{ELM(i,j)}=h(x_i)\cdot h(x_j)=K(x_i,x_j) \end{cases} \tag{9}$$

where x_i and x_j are sample input vectors, and $K(x_i, x_j)$ indicates the kernel functions. When Gaussian kernel functions are used:

$$K(x_i, x_j) = \exp(-\frac{\|x_i - x_j\|^2}{\gamma^2}) \tag{10}$$

where γ is the nuclear parameter. According to the KELM formula above, the weight β of the connection between the output function and the hidden and output layers is:

$$\begin{cases} y(x) = \begin{bmatrix} K(x, x_1) \\ \vdots \\ K(x, x_N) \end{bmatrix}^T (1/C + \Omega_{ELM})^{-1} T \\ \beta = (1/C + \Omega_{ELM})^{-1} T \end{cases} \tag{11}$$

2.3. KPLS Kernel Partial Least Squares Regression

The essential function of PLS is dealing with linear problems. When there is a nonlinear relationship between data, the PLS method is generally not adopted. The KPLS method can solve nonlinear problems by selecting the kernel function.

Assuming a nonlinear mapping $\phi : xi \in R^m \to \phi(xi) \in H$ from original spatial variables $\{xi\}_{i=1}^n$ to a feature space H, Rosipal and Trejo [23] used the relationship between the reproducing kernel Hilbert space (RKHS) and the feature space and developed the linear PLS method into KPLS. The KPLS algorithm can be expressed as follows.

The realization process of KPLS algorithm according to Rosipal and Trejo.

1. Randomly set the initial value u (u can be set to be equal to any column in the Y matrix)
2. According to $w_i = \phi_i^T / ||\phi_i^T u_i||$, calculate the weight vector w_i
3. Calculate the score vector t_i according to the formulas $t_i = \phi_i w_i = \phi_i \phi_i^T u_i / \sqrt{u_i^T \phi_i \phi_i^T}$ and

$\phi_i \phi_i^T u_i / \sqrt{u_i^T \phi_i \phi_i^T} = K_i K_i / \sqrt{u_i^T K_i u_i}$. Unify the vector t_i with the formula $t_i / ||t_i|| \to t_i$
4. According to $q_i = Y_i t_i / ||t_i^T t_i||$ and $u_i = Y_i q_i / (q_i^T q_i)$, calculate the feature vectors
5. Repeat steps 2–5 until convergence
6. Calculate the matrices K and Y as

$$\begin{cases} K_{i+1} = (I - \frac{t_i t_i^T}{t_i^T t_i}) K_i (I - \frac{t_i t_i^T}{t_i^T t_i}) \\ Y_{i+1} = (I - \frac{t_i t_i^T}{t_i^T t_i}) Y_i \end{cases} \tag{12}$$

7. Repeat steps 2–7 until all feature vectors have been calculated

---

3. KPLS Optimal Feature Selection and KELM Fault Diagnosis Method

The specific steps of the fault diagnosis method based on multi-feature extraction and the improved KELM proposed in this paper are as follows. The flow chart is shown in Figure 2.

The realization process of multi-feature and improved KELM fault diagnosis

1. Collect vibration signals of various states of parts
2. Divide the collected vibration signal data, divide the non-overlapping samples into 60 segments, and extract fault features from each segment
3. Construct a high-dimensional feature space and extract 16 time-domain features, 13 frequency-domain features, 8 wavelet energy features, and 8 wavelet packet energy entropy features
4. Input the obtained high-dimensional features into the KELM model for training and testing. Fifty percent are selected as training samples and fifty percent as test samples
5. Adopt a multi-domain KELM fault diagnosis model to identify the fault information

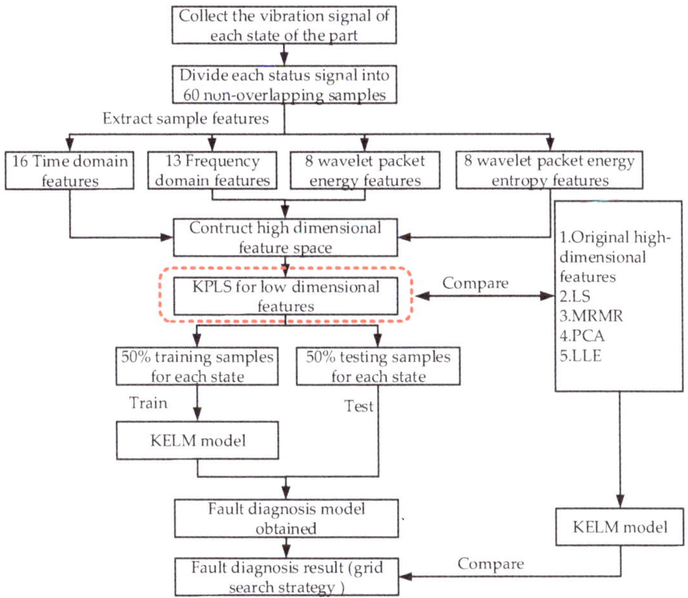

Figure 2. The realization process of KPLS-KELM flow chart.

The advantages and disadvantages of different classification methods are shown in Table 2.

Table 2. The advantages and disadvantages of the KELM and ELM.

	Advantages	Disadvantages
ELM	1. Fast 2. Small sample	1. Low generalization ability
KELM	1. High generalization ability 2. Introduction of kernel function to deal with multi-classification problems 3. Faster calculation speed	

The advantages and disadvantages of the partial least square (LS), max-relevance and min-redundancy (MRMR), principal component analysis (PCA), and locally linear embedding (LLE) dimensionality reduction methods are shown in Table 3.

Table 3. The advantages and disadvantages of dimensionality reduction methods.

	Advantages	Disadvantages
LS	1. Simple 2. Linear	1. Overfitting
MRMR	1. Feature selection based on maximum statistical dependency criteria	1. Underestimates the usefulness of features
PCA	1. Simple and quick 2. Linear methods	1.Difficult to find the right solution
LLE	1. Maintains local linear relationship of samples 2. Low computational complexity	1. Sensitive to the selection of nearest neighbor sample number 2. The manifold learned by the LLE algorithm can only be non-closed
KPLS	1. Introduces kernel function to solve nonlinear problems	

4. Experimental Simulation and Analysis

4.1. Analysis of Test Data

In this part of the study, we used bearing data from Case Western Reserve University in the United States for bearing fault diagnosis [24] to verify the effectiveness of the method proposed in this paper and the effectiveness of the multi-domain feature extraction. All the experiment were implemented using MATLAB 2018A and run on the same Windows 10 machine with an Intel(R) Core (TM) I9-9880h, 2.30GHz CPU and 16GB RAM.

The ten fault states in the rolling bearing fault diagnosis were artificially added, as shown in Table 4, and the three fault diameters of the inner ring, the outer ring, and the rolling element were 0.07 ft, 0.014 ft, and 0.021 ft, respectively. Their time-domain waveform is shown in Figure 3 below.

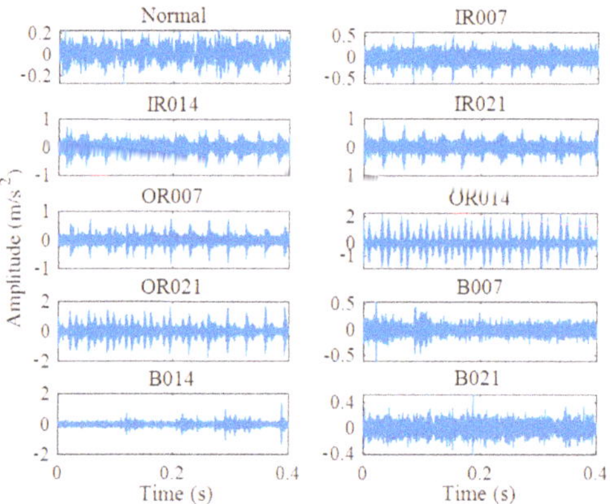

Figure 3. Time-domain diagrams of bearing in ten states (1).

We used the vibration signal of the fan terminal bearing at the motor speed of 1797 r/min as the experimental data, with the data description as shown in Table 4.

Table 4. Experimental bearing sample attributes.

Outer Ring (ft)	Inner Ring (ft)	Rolling Element (ft)
0.014	0.07	0.021

As can be seen from Figure 3, IR007, IR014, IR021, OR007, OR014, OR021, and other signals demonstrated periodic impacts, while the signals of B007, B014, and B021 showed no obvious periodic hints. No amplitude difference between the signals was apparent and neither were the characteristics of the impact nor its period. Hence, the proposed mixed domain could identify the bearing fault type and the degree of the fault.

For the multi-domain feature extraction method proposed in this paper, we divided the vibration signals of each state into 60 non-overlapping samples with a length of 2000 and extracted multi-domain features from each sample separately.

There were thus 16 types of time-domain features and 13 types of frequency-domain features, as shown in Table 1. For the 16 time-frequency domain features, the wavelet packet was decomposed using the wavelet type db5, and the number of decomposed layers was 3. Then, the energy and energy entropy features were extracted from the components obtained by each layer of the wavelet packets. The energy entropy features of the wavelet packets in the time-frequency domain were obtained. The obtained feature is shown in the box diagram in Figure 4.

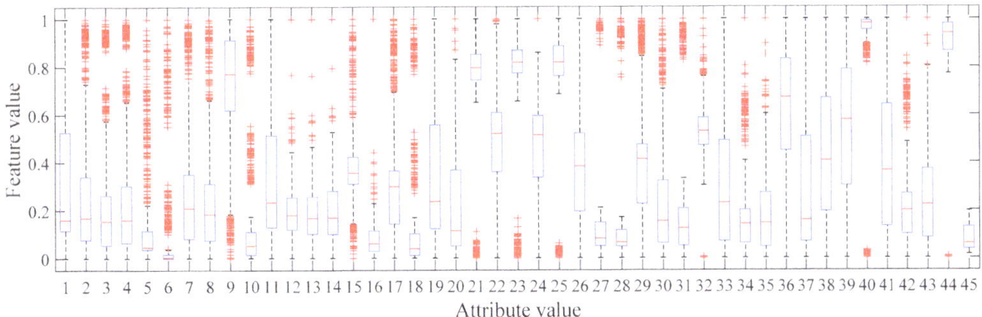

Figure 4. Multi-domain characteristic box plot of bearing in ten states.

After being standardized, most of these features were distributed in the range of 0.2 to 0.8. After the feature extraction, we input the extracted 45 dimension, high-dimensional features into the KELM for fault diagnosis. To verify the effectiveness of the proposed multi-domain features, we extracted 16 types of time-domain features, 13 types of frequency-domain features, 16 types of time-frequency-domain features, and feature sets of the 45 multi-domain features after the dimensionality reduction by KPLS. Due to complex operation conditions, algorithms have difficulty characterizing the fault state completely with its single feature. The combined or multi-domain features could reveal the state information, but there was much redundant information that reduced the accuracy and efficiency of the diagnostic model, so extracting features with moderate dimensions and high sensitivity to each state is the key. LS and MRMR select features that are sensitive to faults or have large contributions, while in feature selection methods such as the Pearson correlation coefficient, distance criterion, and information gain, the emphasis is on the physical meaning of the original features remaining unchanged. PCA employs feature dimension reduction to obtain a compact manifold structure based on feature mapping or feature fusion, and it can extract nonlinear features with the variance in global distribution information unchanged, but it cannot maintain local manifold information, so it is only suitable for linear dimension reduction. The kernel function in KPCA maps data to high dimensional space to obtain nonlinear principal components with higher separability, but the kernel function has a great influence on the results. LLE determines the similarity of neighborhood points using the Euclidean distance, but ignoring the relationship between data leads to unreasonable neighborhood construction.

When comparing the method proposed in this paper with other dimensionality reduction methods, the results shown in Figure 5 were obtained.

The data used for training the KELM model before the dimensionality reduction can be seen in Table 3, including the ten fault states from Figure 3. The total number of samples was 600×45, and each sample contained 45 feature points, comprising 60% of the data used for training. Thus, the number of training sets was 360×45, and the number of testing sets was 240×45. The training accuracy was 100%, and the testing accuracy was 91.67%. The data used for the KELM are shown in Table 5.

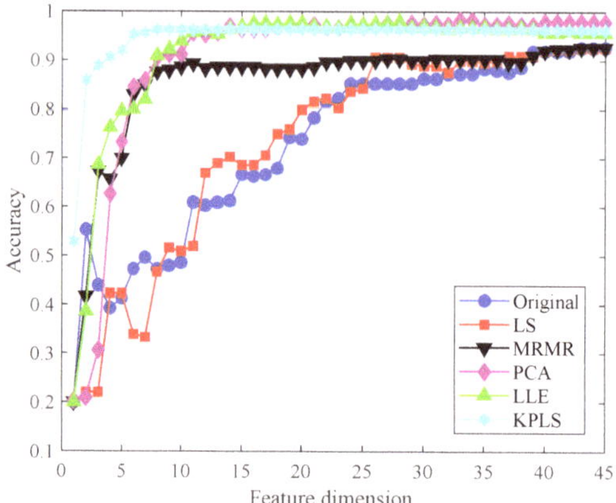

Figure 5. The bearing fault diagnosis accuracy for the KELM with different dimensionality reduction methods.

Table 5. The data used for the KELM.

Total Samples (Number)	Training Sets (Number)	Testing Sets (Number)	Learn Time (s)	Training Accuracy	Testing Accuracy
600×45	360×45	240×45	0.0051	100%	91.67%

We input these feature sets into KELM for fault diagnosis analysis, and the classification results are shown in Figure 6.

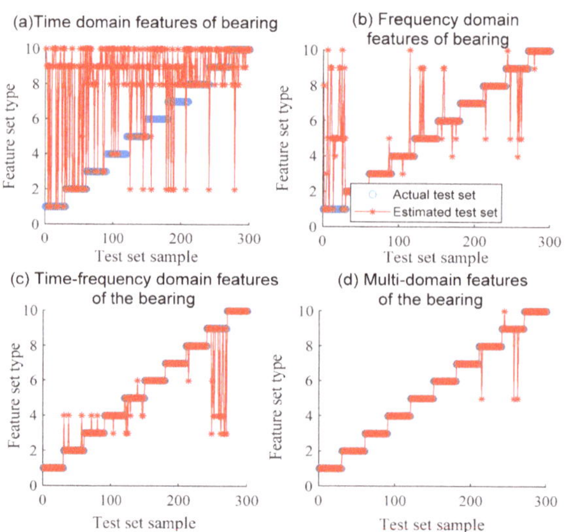

Figure 6. Comparison of diagnoses of the bearing's time-domain, frequency-domain, time-frequency-domain, and multi-domain features.

Comparing the four graphs in Figure 6, it can be seen that the diagnostic results for the multi-domain features of the bearing in Figure 6d are better than the time-frequency-domain features in Figure 6c. Figure 6c shows a better diagnostic result than the frequency domain feature from Figure 6b and the time-frequency domain feature in Figure 6a, while Figure 6b displays a better diagnostic result than Figure 6a. The results shown in Table 6 indicate that the proposed multi-domain feature extraction was better than the time-domain, frequency-domain, and time-frequency-domain features and could achieve satisfactory diagnosis results in the experiment on the bearings with ten fault states. By analyzing in detail the diagnostic results of Figure 6d, we can see that one sample in the third type of rolling element with 007 states was wrongly classified as the fifth type of the outer ring with 007 states. Two samples for the ninth type of the rolling element B0014 were improperly classified as the fifth outer ring with OR007 states. One sample was also wrongly classified into the tenth state of the rolling element with 021 states. There were four wrongly classified samples in total. The accuracy rates for the time domain, frequency domain, time and frequency domain, and multi-feature domain using KPLS were 30.00%, 86.67%, 91.00%, and 97.33%, respectively. As a combination of time-domain, frequency-domain, and time-frequency-domain features, multi-domain features can characterize a fault state fully. The results show that the extraction of 45 multi-domain fault features proposed in this paper had the best effect in bearing fault diagnosis.

Table 6. Diagnosis results under different feature sets.

Feature Set	Diagnostic Time (Seconds)	Accuracy (%)
Time domain	0.0035	30.00
Frequency domain	0.0055	86.67
Time-frequency domain	0.0037	91.00
Multi-feature via KPLS	0.0039	97.33

To eliminate contingency, and as the KELM diagnosis results are often affected by node parameters, we conducted another bearing fault diagnosis experiment with different hidden layer node numbers using the feature extraction method above: time domain, frequency domain, and time-frequency multi-feature domain. The results are shown in Figure 7.

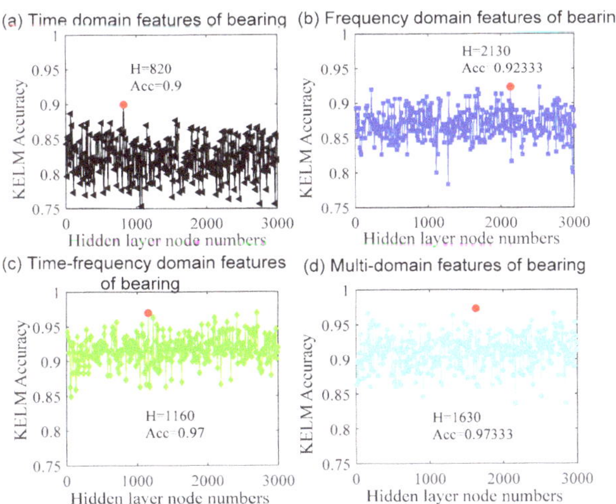

Figure 7. Diagnosis results for the time-domain, frequency-domain, time-frequency domain, and multi-domain features of the bearing with different hidden layers in KELM.

As shown in Figure 7, with the extraction of multi-domain features, while the KELM hidden layer node changed from 0 to 3000 (Table 7), the average fault diagnosis accuracy was maintained at more than 90%. When the hidden layer node value H = 1630, the fault diagnosis accuracy was 97.333%. The results show that the proposed multi-domain feature extraction method had the best fault diagnosis accuracy. It was superior to the time-domain, frequency-domain, and time-frequency-domain feature extraction methods, no matter how the hidden layer nodes changed. At the same time, as shown in Figure 7a, when extracting time-domain features, the average diagnostic accuracy was close to 80%.

Table 7. Corresponding results for different nodes of hidden layer H.

H	Number of Nodes in Hidden Layer H						
	500	820	1000	1500	1630	2130	3000
Time domain	47.65%	58.50%	70.45%	92.22%	91.66%	89.55%	91.66%
Frequency domain	45.57%	67.89%	74.76%	86.56%	83.67%	88.90%	86.43%
Time-frequency domain	65.44%	69.90%	74.34%	84.38%	80.23%	87.77%	85.57%
Multi-domain via KPLS	96.88%	93.45%	92.75%	96.88%	97.33%	93.56%	95.53%

When extracting frequency-domain features, the average diagnostic accuracy was close to 85%. When extracting time-frequency-domain features, the average diagnostic accuracy was close to 90%. When adopting the multi-domain feature proposed, the diagnostic accuracy was higher than or equal to 90%, with the maximum diagnostic accuracy being 97.333%. In this scenario, the number of hidden layer nodes was 1630. The experiment result indicates that the proposed multi-domain feature extraction method had the best diagnostic accuracy.

4.2. Data Analysis of Diaphragm Pump Check Valve

Figure 8a,b show the sensors that were fixed on the shells of the inlet and outlet valves. For each valve, there was one acceleration sensor of the type PCB352C33 (sensitivity: 100 mV/g) and one sound pressure sensor of the type MP021 (50 mV/Pa), respectively. The acceleration sensor collected the shell vibration signal along the Z-axis using three channels, while the sound pressure sensor collected the sound signal along the Y-axis direction.

(a) Inlet valve measuring point

(b) Outlet valve measuring point

Figure 8. Acceleration sensor positions.

Figure 9 shows the vibration signal acquisition device for the check valve. The eight-channel analog signal was amplified, filtered, and converted into A/D by the data acquisition card and sent to the PS PXI-3050EXT 2.7 ghz controller; then, the signal was transferred to the PS PXIE-9108Ext eight-slot industrial computer and stored in the hard disk. When the diaphragm pump check valve ran normally (first 500 h), eight-channel data were collected at the sampling rate of 2560 Hz every 1 h. When the check valve was potentially damaged (500 h~1000 h), eight-channel data were collected every 10 min. When the one-way valve potentially underwent serious failure or damage (after 1000 h), we collected eight-channel data every 2 min.

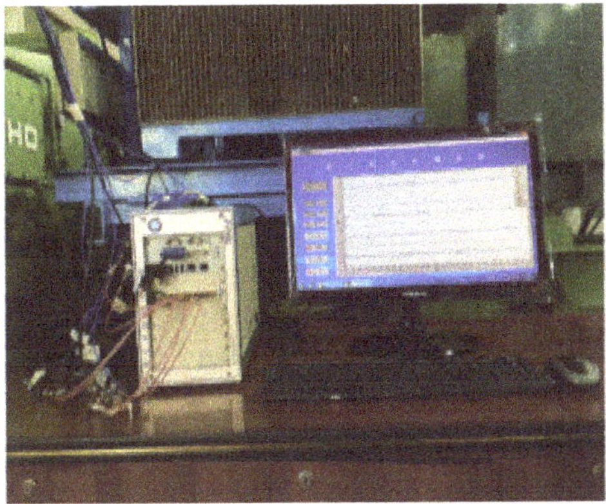

Figure 9. Data acquisition device.

Due to reasons of safety and cost, specific experiments could not be carried out; the damage states of the check valve depended on the actual working conditions. After the check valve was replaced by technicians who work on the site, we checked the damage and recorded the basic fault size and location. The typical damage is shown in Figure 10; Figure 10a shows a stuck valve fault, Figure 10b shows a wear fault, and Figure 10c shows a worn valve fault.

(a) Stuck Valve fault   (b) Wear fault   (c) Worn valve seat   (d)Replaced check valve

Figure 10. Fault check valve.

When the check valve was in a normal state, stuck valve failure state, or wear break-down failure state, we randomly generated a set of time-domain diagrams of vibration signals (as shown in Figure 11). It was found that there were some fault impulses in the middle of the time-domain graph. However, since the impulse period was not apparent, as the noise in the local waveform was inevitable, it was difficult to analyze the type and the cause of failure based on the time-domain waveform alone.

Therefore, the following experiment was used for the vibration signal sample and the multi-domain feature extraction method employed to perform fault diagnosis on the check valve. First, the vibration signal of the check valve in each state was divided into 60 non-overlapping samples, and the number of data points in each sample was 1280. For each non-overlapping piece, we extracted 45 multi-domain features, of which samples 1 to 16 were time-domain features, samples 17 to 29 were frequency-domain features, and samples 30 to 45 were time-frequency-domain features. Among the time-frequency-domain features, we extracted the energy features for the first eight time-frequency-domain features and the energy entropy features for the second eight time-frequency domain features.

The results for the multi-domain features are shown in the boxplot in Figure 12. After normalization, most of the 45 characteristics of the check valve were distributed in the range from 0 to 0.2. Some samples were distributed in the range from 0.2 to 0.8, and very few were distributed in the range from 0.8 to 1.

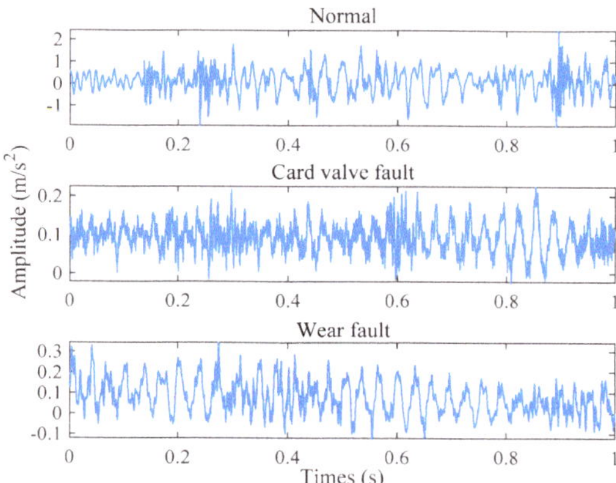

Figure 11. Time-domain diagram of three states of one-way valve.

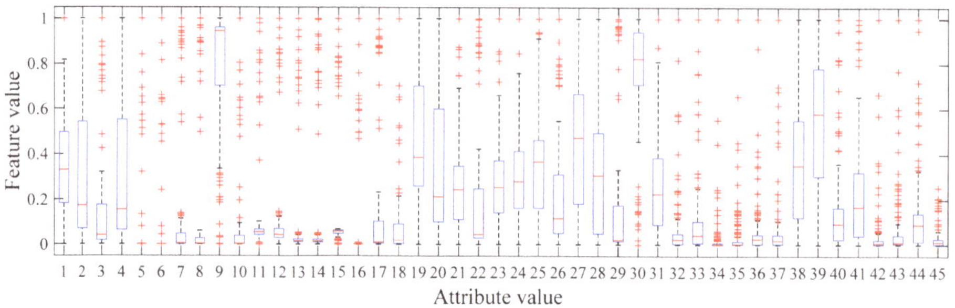

Figure 12. Box diagram of three states for multi-domain features from check valve.

As the number of dimensions of the multi-domain features reached 45, we adopted the KPLS method to reduce the dimensionality of the multi-domain features.

After the KPLS dimensionality reduction, the diagnosis accuracy rate reached more than 95% with only eight dimensions. Figure 13 shows the accuracy results for all the selected dimensionality reduction methods: LS, MRMR, PCA, and LLE.

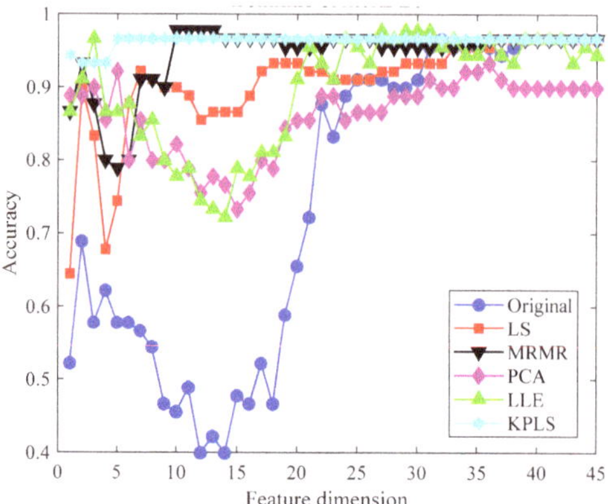

Figure 13. Check valve fault diagnosis accuracy obtained by KELM with different dimensionality reduction methods.

The data used for training the KELM model before the dimensionality reduction can be seen in Table 8. We can see three fault categories in Figure 11; the total number of samples was 180 × 45, each dataset contained 45 feature points, and 60% of the data were used for training. Thus, the number of training sets was 108 × 45, the number of testing sets was 72 × 45, the training accuracy was 100%, and the testing accuracy was 86.11%.

Table 8. The data used for training the KELM.

Total Samples (Number)	Training Sets (Number)	Testing Sets (Number)	Learn Time (s)	Training Accuracy	Testing Accuracy
180 × 45	108 × 45	72 × 45	0.0014	100%	86.11%

To verify the effectiveness of the multi-domain feature extraction results for the check valve, we compared the fault diagnosis results with the time-domain feature, frequency feature, and time-frequency-domain features. The results obtained by the KELM are shown in Figure 14.

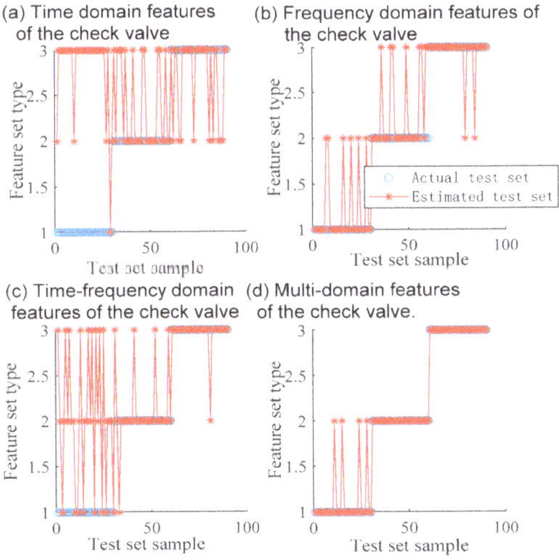

Figure 14. Comparison of diagnostic results for the time-domain, frequency-domain, time-frequency-domain, and multi-domain features of the check valve.

As shown in Figure 14d, four regular sample points were wrongly classified as type 2 (stuck valve failure state). When the number of hidden layer nodes was 200, the final fault diagnosis result was 96.88%.

Compared with the time-domain features, the frequency-domain features, and the time-frequency-domain features for the check valve—which demonstrated accuracies of 45.56%, 82.22%, and 68.89%, respectively—the fault diagnosis results obtained with the multi-domain features, as shown in the Figure 14d, were improved to 96.88%, raising the accuracy rate by 51.32%, 14.66%, and 27.99%. The results shown in Table 9 indicate that the proposed multi-domain feature extraction achieved the optimal diagnosis result in the check valve fault diagnosis experiment, which proves the proposed method's effectiveness.

Table 9. Diagnosis results with different feature sets.

Feature Set	Diagnostic Time (s)	Accuracy (%)
Time domain	0.0039	45.56
Frequency domain	0.0041	82.22
Time frequency domain	0.0039	68.89
KPLS multi-domain feature	0.0047	96.88

As the fault diagnosis of the KELM was affected by the number of hidden layer nodes, we compared the fault diagnosis results of the check valve using KELM when the hidden layer nodes were 0 to 3000, as shown in Figure 15 and Table 10 below.

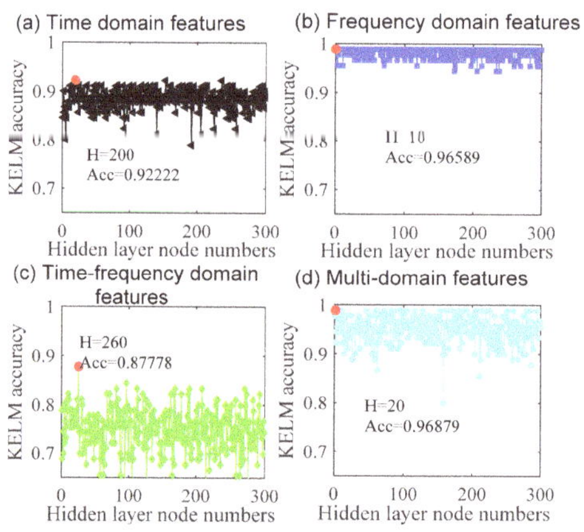

Figure 15. Diagnosis results for check valve with the time-domain, frequency-domain, time-frequency-domain, and multi-domain features using different numbers of hidden layers in KELM.

Table 10. Corresponding results for node numbers different numbers of hidden layers.

	Number of Nodes in Hidden Layer H						
H	10	20	100	200	250	260	300
Time domain	49.66%	57.89%	82.22%	92.22%	85.53%	91.22%	90.32%
Frequency domain	96.59%	94.83%	95.55%	94.78%	93.68%	94.45%	96.20%
Time-frequency domain	65.45%	69.89%	72.32%	74.94%	83.83%	87.77%	87.85%
Multi-domain	92.45%	96.88%	95.56%	96.88%	93.45%	96.50%	96.23%

As shown in Figure 15d, no matter how the hidden layer nodes change, the average diagnosis result was close to 95%, with the maximum diagnosis accuracy being 96.879%. In this scenario, the number of nodes in the hidden layer was ten, which was better than the fault diagnosis results for the time domain, frequency domain, and time-frequency domain. Therefore, the multi-domain feature extraction method proposed in this paper achieved better results in bearing and check valve fault diagnosis, proving the method's effectiveness.

After analyzing different feature sets, we used different classification algorithms to conduct an experimental analysis of multi-domain feature sets. The results are shown in Figure 16 below. It can be seen that the overall accuracy of the proposed KELM exceeded 90%, which was significantly better than the back propagation neural network (BPNN) and ELM.

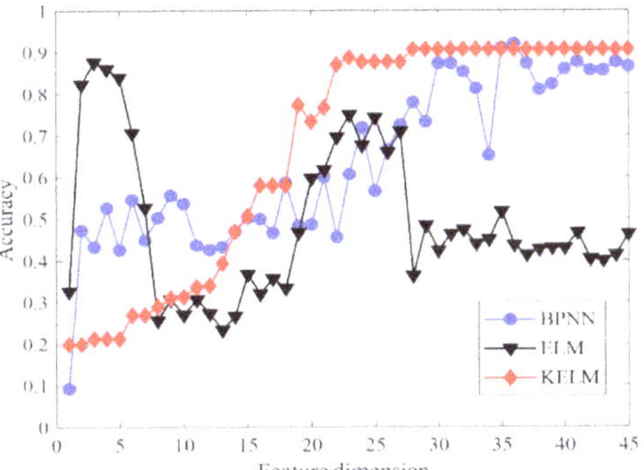

Figure 16. Diagnostic accuracy obtained with KPLS dimensionality reduction features.

5. Conclusions

Investigating the problem of a check valve's fault state being difficult to classify, this paper proposed a diagnosis method based on multi-domain features and KELM. It adopted the method to analyze bearing test fault data and check valve fault data. The conclusions are as follows:

1. When the time-domain, frequency-domain, and time-frequency-domain features were used alone for bearing fault diagnosis, the diagnostic accuracies were 30.00%, 86.67%, and 91.00%, respectively. With the multi-domain feature extraction method after KPLS dimensionality reduction, the accuracy was improved to 97.33%;
2. When the bearing fault diagnosis test was carried out with different numbers of hidden layer nodes, the accuracy was increased from 15.56%, 82.22%, and 68.89% to 97.33% with multi-domain features and KPLS;
3. The proposed KPLS-KELM algorithm could accurately and effectively extract the fault information for the check valve, and the accuracy reached 95%. Compared with the ELM method, KELM is superior for the traditional time-domain, frequency-domain, and time-frequency-domain analysis methods and has higher accuracy.

The accuracy of KELM is affected by the kernel parameters and penalty coefficients. Achieving fast and accurate parameters for different objects is the focus of future research. Application of the theory in practice would be the ultimate end of this research, and online fault diagnosis and big data analysis also need to be considered next.

Though the proposed method achieves better diagnosis results and provides superior accuracy for feature extraction and better robustness in fault diagnosis, the stability of the multi-feature method with coarse-grained data could be further improved. Therefore, it needs more time for training and classification. Next, we will develop a fast training algorithm to facilitate model training.

Author Contributions: Conceptualization, X.Y. and Y.F.; methodology, X.Y.; software, X.Y. and G.Z.; validation, Y.F. and X.W.; formal analysis, C.Z.; investigation, X.Y.; resources, X.Y.; data curation, X.Y.; writing—original draft preparation, Y.F.; writing—review and editing, X.Y.; visualization, X.Y.; supervision, X.W. and C.Z.; project administration, X.Y.; funding acquisition, X.Y. All authors have read and agreed to the published version of the manuscript.

Funding: This research was supported by the Open Research Project of the State Key Laboratory of Industrial Control Technology, Zhejiang University, China (no. ICT2022B06), the PhD research startup foundation of Yunnan Normal University (no. 010000205020503131), and the Fundamental Research Program of Yunnan Province (no. 202101AU070061), Project of Educational Commission of Yunnan Province of China.

Institutional Review Board Statement: Not applicable.

Informed Consent Statement: Informed consent was obtained from all subjects involved in the study.

Data Availability Statement: The data used to support the findings of this study are available from the corresponding author upon request.

Conflicts of Interest: The authors declare no conflict of interest.

References

1. Chen, Z.; Gryllias, K.; Li, W. Mechanical fault diagnosis using Convolutional Neural Networks and Extreme Learning Machine. *Mech. Syst. Signal Process.* **2019**, *133*, 106272. [CrossRef]
2. Peng, Z.; Cheng, L.; Zhan, J.; Yao, Q. Fault classification method for rolling bearings based on the mlllti-feature extraction and modified Mahalanobis-Taguchi system. *J. Vib. Shock* **2020**, *39*, 249–256.
3. Jiang, J.; Hu, Y.; Ke, Y. Fault diagnosis of diesel engines based on wavelet packet energy spectrum feature extraction and fuzzy entropy feature selection. *J. Vib. Shock* **2020**, *39*, 273–277.
4. Ding, T.; Yan, Y.; Li, X.; Liu, L. Study of scintillation detector fault diagnosis based on ELM method. *Nucl. Instrum. Methods Phys. Res. Sect. A Accel. Spectrometers Detect. Assoc. Equip.* **2022**, *1032*, 166637. [CrossRef]
5. Lee, J.; Sun, Z.; Tan, T.B.; Mendez, J.; Flores-Cerrillo, J.; Wang, J.; He, Q.P. Remaining Useful Life Estimation for Ball Bearings Using Feature Engineering and Extreme Learning Machine. *IFAC-PapersOnLine* **2022**, *55*, 198–203. [CrossRef]
6. Pang, S.; Yang, X.; Zhang, X.; Lin, X. Fault diagnosis of rotating machinery with ensemble kernel extreme learning machine based on fused multi-domain features. *ISA Trans.* **2020**, *98*, 320–337. [CrossRef]
7. Wang, X.B.; Zhang, X.; Li, Z. Ensemble extreme learning machines for compound-fault diagnosis of rotating machinery. *Knowl.-Based Syst.* **2019**, *188*, 105012. [CrossRef]
8. Yin, G.; Zhang, Y.T.; Li, Z.N.; Ren, G.Q.; Fan, H.B. Online fault diagnosis method based on incremental support vector data description and extreme learning machine with incremental output structure. *Neurocomputing* **2014**, *128*, 224–231. [CrossRef]
9. Harishvijey, A.; Raja, J.B. Automated technique for EEG signal processing to detect seizure with optimized Variable Gaussian Filter and Fuzzy RBFELM classifier. *Biomed. Signal Process. Control* **2022**, *74*, 103450. [CrossRef]
10. Shen, J.; Xu, F. Method of fault feature selection and fusion based on poll mode and optimized weighted KPCA for bearings. *Measurement* **2022**, *194*, 110950. [CrossRef]
11. Li, Z.; Wang, B.; Zhu, B.; Wang, Q.; Zhu, W. Thermal error modeling of electrical spindle based on optimized ELM with marine predator algorithm. *Case Stud. Therm. Eng.* **2022**, *38*, 102326. [CrossRef]
12. Huang, G.B. An insight into extreme learning machines: Random neurons, random features and kernels. *Cogn. Comput.* **2014**, *6*, 376–390. [CrossRef]
13. Huang, G.; Huang, G.B.; Song, S.; You, K. Trends in extreme learning machines: A review. *Neural Netw.* **2015**, *61*, 32–48. [CrossRef] [PubMed]
14. Chen, Q.; Wei, H.; Rashid, M.; Cai, Z. Kernel extreme learning machine based hierarchical machine learning for multi-type and concurrent fault diagnosis. *Measurement* **2021**, *184*, 109923. [CrossRef]
15. Su, H.; Xiang, L.; Hu, A.; Gao, B.; Yang, X. A novel hybrid method based on KELM with SAPSO for fault diagnosis of rolling bearing under variable operating conditions. *Measurement* **2021**, *177*, 109276. [CrossRef]
16. Kang, S.; Ma, D.; Wang, Y.; Lan, C.; Chen, Q.; Mikulovich, V. Method of assessing the state of a rolling bearing based on the relative compensation distance of multiple-domain features and locally linear embedding. *Mech. Syst. Signal Process.* **2017**, *86*, 40–57. [CrossRef]
17. Xue, X.; Zhou, J. A hybrid fault diagnosis approach based on mixed-domain state features for rotating machinery. *Isa Trans.* **2016**, *66*, 284–295. [CrossRef]
18. Attoui, I.; Fergani, N. A New Time-frequency Method for the Identification and classification of Boutasseta faults. *J. Jilin Univ. (Nat. Sci. Ed.) Sound Vib.* **2017**, *397*, 241–265. [CrossRef]
19. Yao, B.; Su, J.; Wu, L.; Guan, Y. Modified Local Linear Embedding Algorithm for Rolling Element Bearing Fault Diagnosis. *Appl. Sci.* **2017**, *7*, 1178. [CrossRef]
20. Chen, R.; Chen, S.; Yang, L.; Wang, J.; Xu, X.; Luo, T. Looseness diagnosis method for connecting bolt of fan foundation based on sensitive mixed-domain features A Mathematical Model for Solving mathematical Congestion. *Comput. Modeling Comput. Modeling* **2005**, *54*, 554.
21. Huang, G.-B.; Zhu, Q.-Y.; Siew, C.-K. Extreme learning machine: Theory and applications. *Neurocomputing* **2006**, *70*, 489–501. [CrossRef]
22. Peng, Y.; Li, Q.; Kong, W.; Qin, F.; Zhang, J.; Cichocki, A. A joint optimization framework to semi-supervised RVFL and ELM networks for efficient data classification. *Appl. Soft Comput.* **2020**, *97*, 106756. [CrossRef]
23. Rosipal, R.; Trejo, L.J.; Wheeler, K. Locally-Based Kernal Pls Smoothing to Non-parametric Regression Curve Fitting. 2002; pp. 1–36. Available online: https://ntrs.nasa.gov/citations/20030015244 (accessed on 23 June 2022).
24. Loparo, K. Bearings Vibration Data Set. Case Western Reserve University. Available online: https://engineering.case.edu/bearingdatacenter/download-data-file (accessed on 23 June 2022).

Article

Design, Optimization and Cutting Performance Evaluation of an Internal Spray Cooling Turning Tool

Leping Liu, Shengrong Shu *, Huimin Li and Xuan Chen

Key Laboratory of Conveyance and Equipment, Ministry of Education, School of Mechatronics & Vehicle Engineering, East China Jiaotong University, Nanchang 330013, China
* Correspondence: shushengrong201@163.com; Tel.: +86-0791-8704-6132

Abstract: The traditional flood cooling method applied in the internal turning process has disadvantages, such as having a low cooling efficiency and being environmentally unfriendly. In the present work, an internal spray cooling turning tool was designed, and the performance was numerically and experimentally accessed. The heat transfer simulation model of the internal spray cooling turning tool was established by ANSYS Fluent, and the tool cooling structure parameters were optimized by the Taguchi method based on the CFD simulations, and obtains the diameters of the upper and lower nozzles of 3 mm and 1.5 mm, respectively; the distance between the upper nozzle and the tool tip of 18.5 mm. To evaluate the cutting and cooling performance of the optimized tool, internal turning experiments were conducted on QT500-7 workpieces. Results show that the optimized tool with internal spray cooling led to lower workpiece surface roughness and chip curling, compared to the conventional tools.

Keywords: internal turning; internal spray cooling; tool design; CFD simulations; green cutting

Citation: Liu, L.; Shu, S.; Li, H.; Chen, X. Design, Optimization and Cutting Performance Evaluation of an Internal Spray Cooling Turning Tool. *Coatings* **2022**, *12*, 1141. https://doi.org/10.3390/coatings12081141

Academic Editor: Diego Martinez-Martinez

Received: 14 July 2022
Accepted: 3 August 2022
Published: 8 August 2022

Publisher's Note: MDPI stays neutral with regard to jurisdictional claims in published maps and institutional affiliations.

1. Introduction

The inner circular surface machining quality of the components, such as engine cylinder, shaft sleeve, hydraulic cylinder and connecting rod, is one of the key factors that significantly affects their service performance. The inner circular surface, represented by two types named through-hole and blind-hole, are mainly machined by turning, drilling or boring processes. The internal turning is widely used to machine the inner circular surface. However, chatter vibration may happen due to the large overhang of the toolholder when turning the inner circular surface, which will affect the machining accuracy. Many methods aiming to restrain this vibration have been developed to increase the inner circular surface quality [1,2]. In addition, the high cutting temperature has a great influence on the machining quality [3]. It is well known that the heat generated in the turning process is mainly transmitted by chips, turning tools and workpieces [4,5]. However, the turning tool operates inside the workpiece when turning the circular surface, and the chips are difficult to evacuate. This causes an extremely high cutting temperature inside the workpiece and accelerates tool wear, and consequently deteriorate the inner circular surface quality. Thus, designing cutting tools, reducing cutting friction and controlling the cutting temperature during the machining process are important when facing the semi-closed heat dissipation space and low thermal conductivity of the workpieces.

Currently, the flood cooling is widely used in the process of cutting temperature reduction. However, this method is known as having low coolant utilization, high pollution levels and processing cost of waste coolant [6,7]. To improve the cooling efficiency and achieve green and environmentally friendly cutting, various cooling methods have been developed and utilized in the cutting process, such as the cryogenic cooling [8], heat pipe cooling [9], closed internal cooling [10], minimum quantity lubrication (MQL) [11], and cryogenic minimum quantity lubrication (CMQL) [12]. Therein, the MQL, or so-called spray cooling can avoid the utilization of complex and expensive equipment of cryogenic cooling, and improve the insufficient cooling and lubrication capacity. Li et al. [13] carried out the sliding tests between YG8 cemented carbide and austempered ductile iron under dry, cold air and the MQL conditions by using a tribometer. Results show that the tool wear rate was the lowest under the MQL cooling condition. Das et al. [14] evaluated the machining performance of the austenitic stainless steel under the dry cutting, compressed air, flood and MQL conditions. They found that the tool life under the MQL conditions was improved compared to other conditions, and the chip separation speeds were also improved.

The external spray cooling can be applied on conventional tools when turning externally, therein having no specific limitation on the cutting space. However, the internal spray cooling can be more efficient and effective in drilling and internal turning. Zeilmann et al. [15] carried out the drilling experiment and found that compared to the external spray cooling, the internal spray cooling significantly reduced the cutting temperature. Jessy et al. [16] tested the cutting performance of the GERP composite material under external and internal cooling conditions. Results show that the internal cooling method reduces the average temperature by 66% compared to the external cooling method. Li et al. [17] reviewed the drilling machinability of CGI under the dry cutting, compressed air and MQL conditions. They summarized that compared to the dry cutting condition, the tool life could be greatly improved under the compressed air and MQL (5 mL/h) conditions.

A well-designed internal cooling cutting tool can precisely send the coolant into the cutting zone to improve the cooling and lubrication efficiency [18–20]. Obikawa et al. [21] conducted CFD simulations on the internal cooling tools with three types of nozzle structures. They found that the cover-type nozzle provided the best performance. Duchosal et al. [22] conducted the optimization on the milling tools and found that the spray angle had significant influence on the distribution of the oil mist. The spray angle of 75° provided the best cooling performance. Zhang et al. [23] compared the milling performance with H13 steel using three types of flow channels in the internal cooling milling tools. Results show that double straight channel was most beneficial to the tool life improvement. Peng et al. [24] designed and optimized an internal cooling external turning tool with microchannel structures. The cutting experiments reveal that compared to the conventional flood cooling and external cooling turning tools, the internal cooling technology led to lower cutting temperature, and improved the surface topography as well.

The aforementioned studies mainly focused on the cooling method design for the external turning tools, drilling tools and milling tools, rather than on the development of the environmentally friendly cooling structure for the internal turning tool. Thus, in this work, an internal spray cooling turning tool structure was proposed based on the requirements on the internal surface turning of the workpiece. The Taguchi orthogonal design method based on CFD simulations were conducted to optimize the cooling structure parameters of internal spray cooling turning tool, and the turning tool was prepared according to the optimization results. A cutting test was carried out on the inner cylinder of the QT500-7 workpiece. The influence of internal spray cooling turning tool parameters on the cutting temperature, workpiece surface roughness and chip morphology were investigated.

2. Cooling Structure Design of the Internal Spray Cooling Turning Tool

Significantly high cutting temperature is often observed on the internal surface of the hollow cylindrical workpiece during the turning process due to the semi-closed space and poor material thermal conductivity, and thus deteriorates the workpiece surface quality and tool life. Compared to the tools with a single nozzle designed on the front or flank face, the internal cooling tools with two spray cooling nozzles can significantly improve the cooling efficiency and tool wear performance [25].

Figure 1 shows the structure of the designed internal spray cooling turning tool. The S25R-MCLNR12 internal turning tool was selected as the prototype with a toolholder diameter of 25 mm. The green part shown in Figure 1 represents the flow channel of the coolant flowing through the tool, the main coolant flow channel is designed in the center of the toolholder, and the cooling nozzles are designed on the rake and flank faces of the tool, respectively. The coolant inlet is connected with the external spraying cooling equipment, which designed at the end of the toolholder. During the turning process, the compressed air carrying a certain amount of coolant flows into the tool via the inlet, passes through the internal cooling channel and then sprays out from the nozzles to lubricate the tool-chip and tool-workpiece interfaces. In addition, the dynamic pressure generated by the compressed air are expected to contribute to the chip removal.

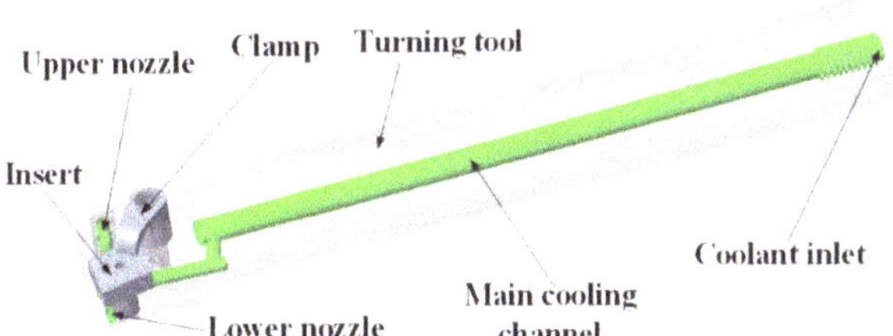

Figure 1. The structure schematic of internal spray cooling turning tool.

In the turning process with spray cooling, the number of nozzles, diameter of the nozzles, distances between nozzles and cutting zone are the key parameters that significantly affecting the cutting temperature variation [26,27]. In Section 3, a numerical simulation model of the internal turning was established by using the ANSYS Fluent to study the influence of the nozzle diameter, the distance between the nozzle and tool tip on the cutting performance. Then the parameters of the tool cooling structure were optimized by Taguchi method based on the CFD simulations.

3. Structure Optimization of Internal Spray Cooling Turning Tool

3.1. Establishment of Simulation Model

The inner surface diameter and length of the workpiece were set as 45 mm and 200 mm, respectively; the outer diameter of the workpiece was set as 70 mm to improve the simulation efficiency. The fluid domain was established inside the tool and workpiece considering the flow channel of the compressed air and coolant droplets. Figure 2 shows the geometrical model of the internal spray cooling turning tool. The small square surface (L1 × L2) on the tool rake face represents the tool-chip interface, which was set as 1.0 mm × 0.5 mm in all simulations. The heat flux was set as 40 W/mm^2 and was applied on this square surface to simulate the cutting heat transfer.

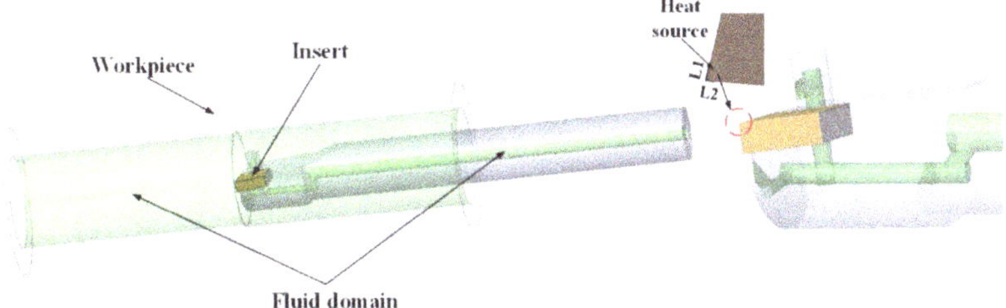

Figure 2. The geometric model of numerical simulation for the internal spray cooling turning tool.

Since the volume fraction of the coolant droplets in the fluid domain is much smaller than air, the discrete phase model (DPM) was chosen to simulate the flow behaviour of the coolant droplets. Considering the evaporation and boiling of the droplets after being sprayed into the cutting zone, the species transport model is essential to be activated.

Because of the entrainment occurring during the spraying and the disturbance due to the high-speed fluid impingement on the irregular surfaces of tool and workpiece, the realizable k-ε turbulence model, highlighted by its efficient prediction on the circular jet and plane jet, was applied to simulate the turbulent flow of fluid [28]. The wall-film model was utilized to simulate the liquid droplets colliding with the surfaces of the tools and workpiece, thus forming thin films, splashing, boiling and vaporization.

As a green cooling method, the amount of coolant used in the spray cooling process is usually within 50–500 mL/h [29]. Here, water was used as the coolant in the spray cooling, and the flow rate of coolant was 50 mL/h. The initial temperature of the cutting was 20 °C in the simulations. Table 1 lists the input parameters of the thermal-fluid-solid coupling simulations for the internal spray cooling turning tool.

Table 1. Simulation parameters.

Variable	Value	Variable	Value
Density of insert (kg/m^3)	14,900	Heat flux on contact area (W/mm^2)	10
Thermal conductivity of insert (W/m·K)	52.3	initial temperature (°C)	20
Specific heat of insert (J/kg·K)	302	Specific heat of water (J/kg·K)	4182
Density of toolholder (kg/m^3)	7850	Viscosity of water (kg/m·s)	0.001
Thermal conductivity of toolholder (W/m·K)	16.3	Density of water (kg/m^3)	998.2
Specific heat of toolholder (J/kg·K)	502	Thermal conductivity of water (W/m·K)	0.6
Tool-chip contact area (L1 mm × L2 mm)	1 × 0.5	Inlet pressure of spray cooling (MPa)	0.3

To improve the simulation efficiency and ensuring the accuracy simultaneously, the grids were meshed densely near the cutting tip and flow channel wall, while were meshed coarsely in other regions. A mesh independence study was carried out based on varying meshed grid sizes of different parts in each design scheme of the Taguchi method. Meanwhile, the CFD simulation was conducted with varying mesh grid sizes. Figure 3 shows that the effect of the grid number on the cutting temperature is insignificant within the range of 2.56–3.2 million. Figure 4 shows the meshed simulation model of the fluid-solid coupling heat transfer during the internal turning with internal spray cooling.

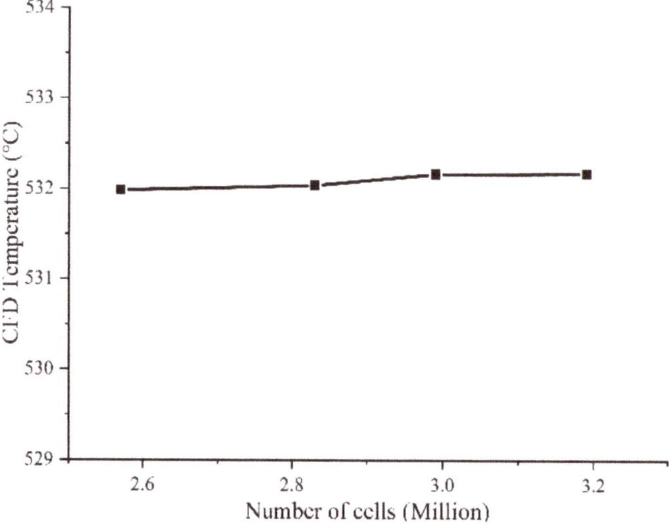

Figure 3. Mesh independence calculating result.

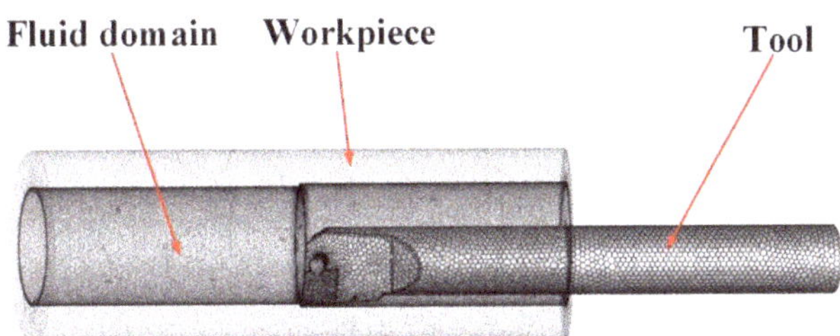

Figure 4. The mesh of simulation model of fluid-solid coupling heat transfer.

3.2. Optimization of Tool Cooling Structure Parameters

The Taguchi orthogonal design method based on the CFD simulations was used to optimize the parameters of the tool cooling structure. Figure 5 shows the structure parameters that to be optimized. The lower nozzle-tip distance (LND) was 8.5 mm in this design. The upper nozzle-tool tip distance (UND), the upper nozzle diameter (UD) and the lower nozzle diameter (LD) were the variables that to be optimized. Note that the influence of the distance from the lower nozzle to tool tip on the cutting temperature was not considered due to the space limitation of the toolholder lower end.

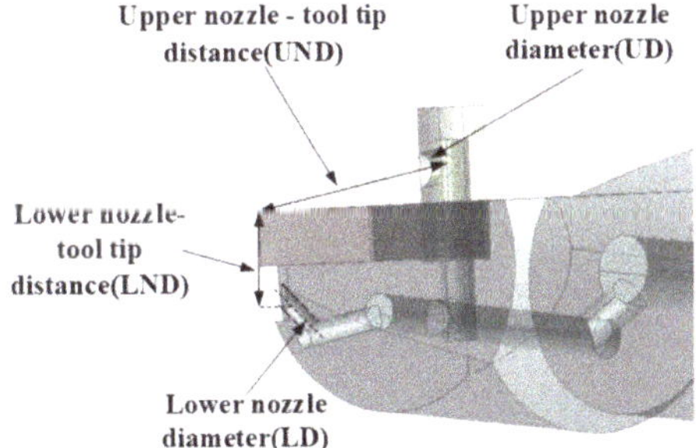

Figure 5. Geometric parameters to be optimized of spray cooling structures.

Considering the influence of the cooling structure layout on the tool rigidity and machinability, each design parameter had three levels as listed in Table 2. The maximum temperature of the tool-chip interface was taken as the output value of the Taguchi test.

Table 2. List of the geometric design parameters and their levels.

Parameters		Levels		
		1	2	3
A	UND (mm)	18.5	21.5	24.5
B	UD (mm)	1	2	3
C	LD (mm)	1	1.5	2

3.3. Simulation Results Analysis

For each design scheme in the orthogonal table $L_9(3^4)$, the maximum temperature of the tool was obtained through the simulations, and the range analysis of the orthogonal test was carried out as shown in Table 3. Therein, the range value of the effects of the upper nozzle diameter, upper nozzle-tool tip distance and lower nozzle diameter on the maximum temperature are 1.61 °C, 1.73 °C and 5.17 °C, respectively, indicating that compared to the upper nozzle diameter and upper nozzle-tool tip distance, the lower nozzle diameter has greater influence on the cooling performance of the internal spray cooling turning tool.

Table 3. Simulation results and range analysis of $L_9(3^4)$ orthogonal test.

No.	Upper Nozzle–Tool Tip Distance (A) /mm	Upper Nozzle Diameter (B) /mm	Lower Nozzle Diameter (C) /mm	Maximum Temperature/°C
1	18.5	1	1	536.19
2	18.5	2	1.5	529.99
3	18.5	3	2	531.47
4	21.5	1	1.5	529.14
5	21.5	2	2	535.29
6	21.5	3	1	534.47
7	24.5	1	2	536.58
8	24.5	2	1	535.12
9	24.5	3	1.5	531.14
K_1	532.55	533.97	535.26	K_i is the average value of
K_2	532.96	533.47	530.09	maximum temperature at
K_3	534.28	532.36	534.44	each factor level; R is
R	1.73	1.61	5.17	the value of range

Figure 6 shows the effect of each parameter on the mean of the maximum temperature. The maximum temperature of the tool-chip contact area increases with the increasing UND (A) from 532.55 °C to 534.28 °C, because the increasing distance between the upper nozzle and tool tip results in the decrease in the coolant delivery to the tip, and thus increases the coolant evaporation efficiency. Therefore, less air and liquid droplets flows into the cutting area, resulting in less convective heat transfer. The temperature of the tool-chip contact area decreases with the increasing UD (B) from 533.97 °C to 532.36 °C. Larger upper nozzle diameter leads to larger flow rate of coolant droplets when the spray pressure is constant, thus more droplets flow into the cutting zone and more heat is transferred. With increasing LD (C), the maximum temperature decreases to 530.09 °C till 1.5 mm and obtains the best cooling performance. Then it increases to 534.44 °C at 2.0 mm. This is mainly because the lower nozzle diameter affects the amount of the droplets ejected from both the lower and upper nozzles.

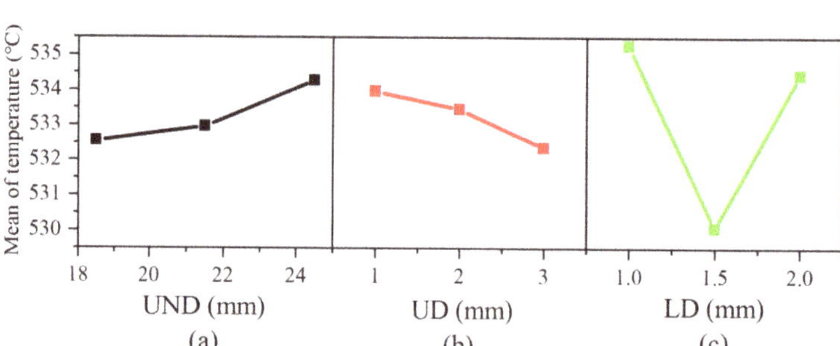

Figure 6. The diagram of main effects of the three variables on maximum temperature. (**a**) Influence of UND to maximum temperature, (**b**) Influence of UD to maximum temperature, (**c**) Influence of LD to maximum temperature.

According to the range analysis results, the optimal combination of the structure parameters is A1, B3, and C2, namely the UND (A) is 18.5 mm, the UD (B) is 3 mm, and the LD (C) is 1.5 mm. The numerical study was conducted for the optimized internal spray cooling turning tool under the same boundary condition. Figure 7a shows the temperature distributions of the tool with the optimal structure parameters combination. The maximum temperature is 528.48 °C at the tool tip, which is lower than the minimum temperature listed in Table 1. This demonstrates that the cooling performance of the tool with optimized structure parameters combination was improved. Figure 7b illustrates the fluid pathline of the internal spray cooling process. The compressed air is sent through the inner channel of the tool and then is sprayed out rapidly from the upper and lower nozzles. The air is sprayed on the tool tip and then diffused rapidly.

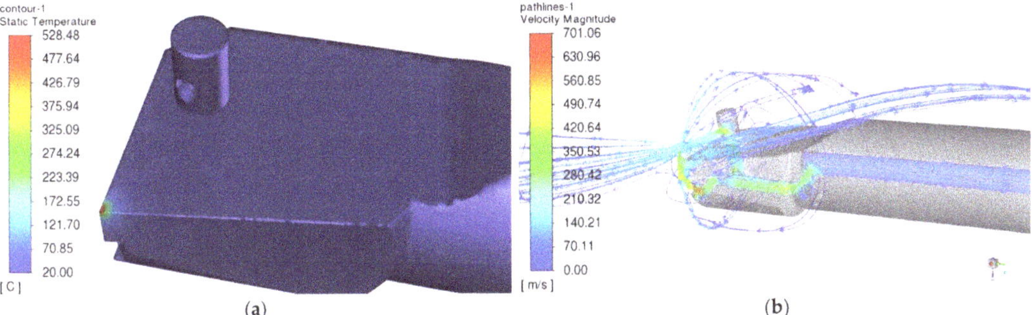

Figure 7. Temperature contour and fluid pathline of the optimized tool. (**a**) Temperature contour of the optimized tool; (**b**) Fuid pathline of the internal spray cooling process.

3.4. Influence of Inlet Pressure on Cooling Performance

Inlet pressure is an important parameter that directly affecting the spray cooling performance [22]. The maximum spray pressure provided by the spray cooling equipment was 0.6 MPa, thus the spray pressure was selected as 0.05 MPa, 0.1 MPa, 0.2 MPa, 0.3 MPa, 0.4 MPa, 0.5 MPa and 0.6 MPa to investigate the influence of the spray pressure on the cutting temperature of the optimized tool.

Figure 8 shows the coolant droplets velocity under different inlet pressures ranging from 0.1 MPa to 0.6 MPa. Overall, with increasing tool inlet pressure from 0.1 MPa to 0.6 MPa, the droplet velocity increases from 367.7 m/s to 861.6 m/s, and the droplet distribution range changes from the local area near the tip to the entire internal surface of the workpiece, which results in more rapid heat transfer.

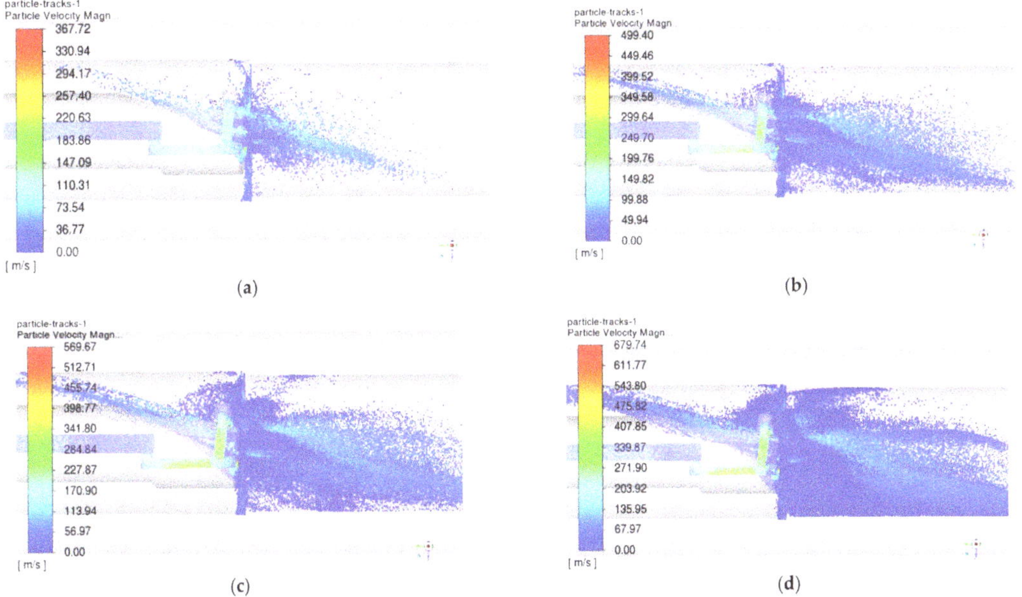

Figure 8. *Cont.*

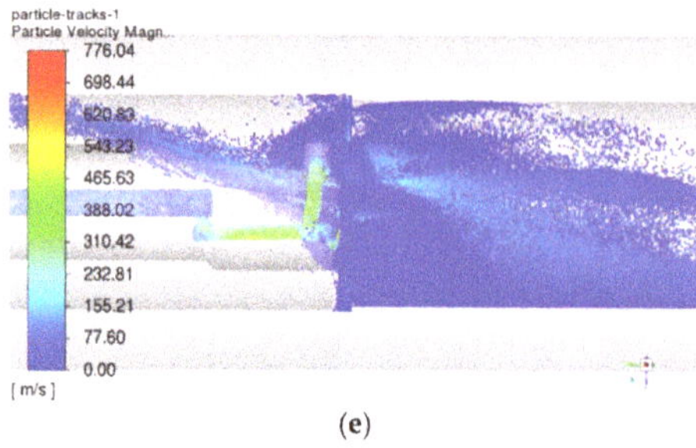

(**e**)

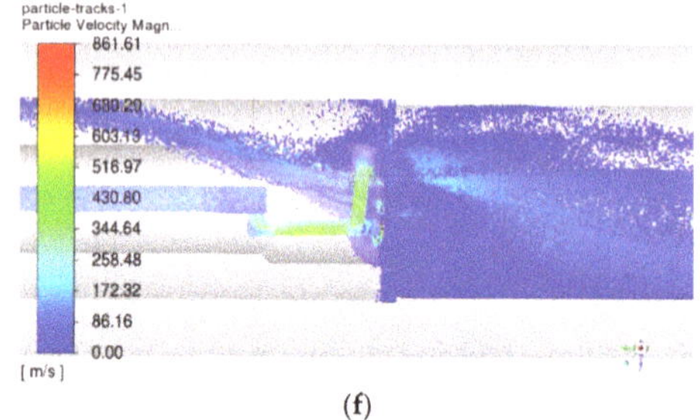

(**f**)

Figure 8. Droplets velocity under different inlet pressures, (**a**) 0.1 MPa, (**b**) 0.2 MPa, (**c**) 0.3 MPa, (**d**) 0.4 MPa, (**e**) 0.5 MPa, (**f**) 0.6 MPa.

Figure 9 shows the influence of tool inlet pressure on the maximum temperature of the tool-chip interface. The maximum temperature first decreases rapidly but then mildly with the increase in spray pressure. This indicates that the spray cooling effectively decreases the cutting temperature even with low inlet pressure such as 0.05 MPa. The high-speed airflow under the spray condition can rapidly facilitate the convective heat transfer, so when the tool inlet pressure is in the low region (<0.1 MPa), the cutting temperature decreasing trend is more significant. Therein, the temperature at 0.05 MPa is approximately 60 °C lower than that of the dry condition. When the inlet pressure is larger than 0.3 MPa, the temperature decreasing tends to be insignificant, which is mainly due to the gradual saturation of the convective heat transfer.

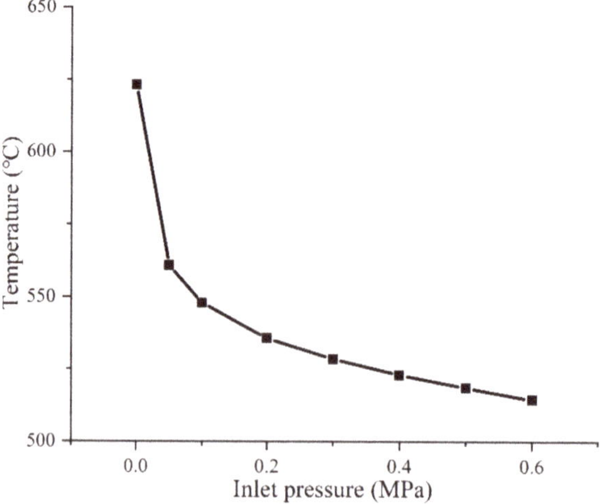

Figure 9. Maximum temperatures of tool under different inlet pressures.

4. Tool Preparation and Cutting Experiments

4.1. Tool Manufacturing

The S25R-MCLNR12 internal turning tool was used as the prototype tool. According to the optimization results, both the internal flow channels of toolholder and the nozzle were manufactured. Figure 10 shows the manufactured internal spray cooling turning tool. The main cutting-edge angle of the tool is 95°, the rake angle is −6°, and the relief angle is 5°. The TiAlN coated VP15TF cemented carbide insert produced by MITSUBISHI (Tokyo, Japan) was adopted. Small holes with a diameter of 0.6 mm were drilled by the electrical discharge machining near the tip of insert for installing thermocouples to measure the cutting temperature.

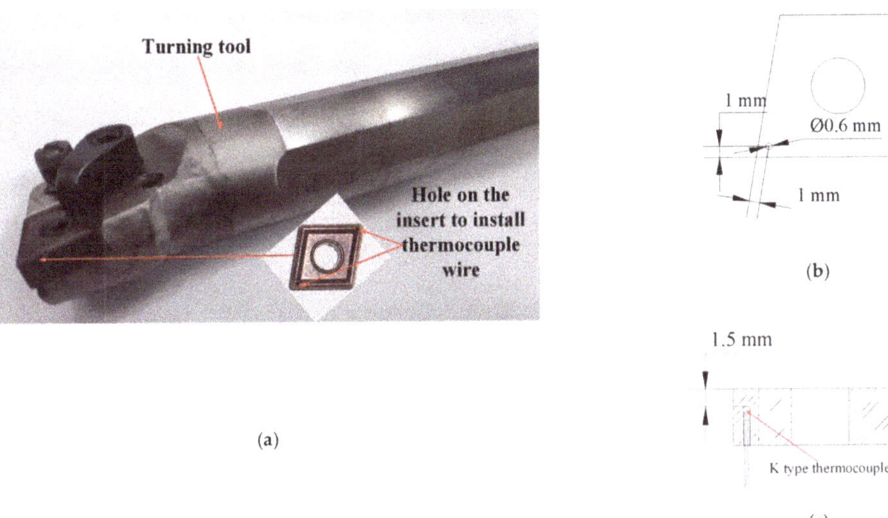

Figure 10. The manufactured internal spray cooling turning tool and thermocouple position. (**a**) the internal spray cooling turning tool, (**b**) diagram of distance between thermocouple and main and minor cutting edges, (**c**) diagram of the distance between the thermocouple and the rake face.

4.2. Experimental Conditions and Settings

Figure 11 shows the internal turning experimental platform, which includes the lathe, spray cooling system, internal spray cooling turning tool and data measuring system. The experiments were conducted on HuaZhong CNC machine tool CK6136B (HOTON, Shandong, China), the maximum rotation diameter of the machine tool is 360 mm, and the maximum spindle speed is 6000 rpm. The workpiece is a hollow cylinder with inner and outer diameters of 45 mm and 100 mm, and its length is 200 mm. The workpiece is made of QT500-7 with a hardness of 170–230 HBS, and the strength is approximately 500 MPa. The coolant used for internal spray cooling is a mixture of water and oil with a volume ratio of 30:1. A K-type thermocouple was embedded into the small hole of the insert to measure the cutting temperature, and the thermocouple was connected with the JK808 temperature tester (Jinko Electronic Technology Co. Ltd., Changzhou, China), which was linked to the computer for collecting the real-time temperature data. The TR200 handheld surface roughness measuring instrument (JiTai Tech Detection Device Co., Ltd., Beijing, China) was used to capture the roughness of internal surface of workpiece. After each cutting test, the surface roughness was measured at six different locations along the circumferential direction of internal surface of workpiece, and the average value was taken as the output value surface roughness. Chips were also collected and the morphology were captured by a handheld microscope. The cutting experiments were carried out under both the dry and internal spray cooling conditions for comparison purposes, and the corresponding operation parameters of the experiment are listed in Table 4.

Table 4. The parameters of cutting experiments.

Parameter	Value
Cutting speed (m/min)	60, 80, 100, 120, 140
Feed rate (mm/r)	0.1
Depth of cut (mm)	0.5
Cooling conditions	0.1–0.6 MPa/dry cutting
Coolant flow rate (mL/h)	50

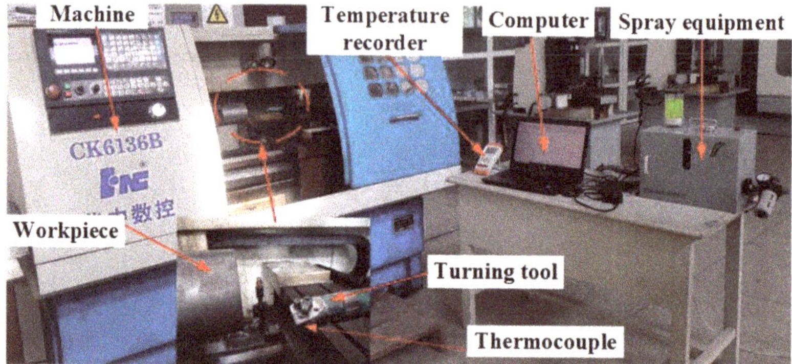

Figure 11. Internal turning experimental platform.

4.3. Results and Discussion

4.3.1. Cutting Temperature

The cutting temperature were measured by the thermocouple inserted in the thermocouple node in the insert as shown in Figure 10b,c. The experimental and CFD temperatures are shown in Figure 12a. By adjusting the heat flux (22 W/m^2 in this case) in the tool-tip contact surface, the errors between the CFD and experimental results could be minimized (5% in this case). In the CFD simulation, the effect of the chip on the cutting temperature was not considered, thus when the inlet pressure is smaller than 0.2 MPa (leading to lower air pressure), the experimental temperature is larger due to the poor chip removal ability. When the spray pressure is below 0.1 MPa, the measured temperature is 7.8 °C and 15.0 °C higher than that of the simulation in 0.1 MPa and dry cutting; when the pray pressure is larger than 0.2 MPa, the CFD temperature can be larger than the measure temperature due to the chip removal. Overall, the effects of the inlet pressure on the CFD and measured cutting temperatures are similar.

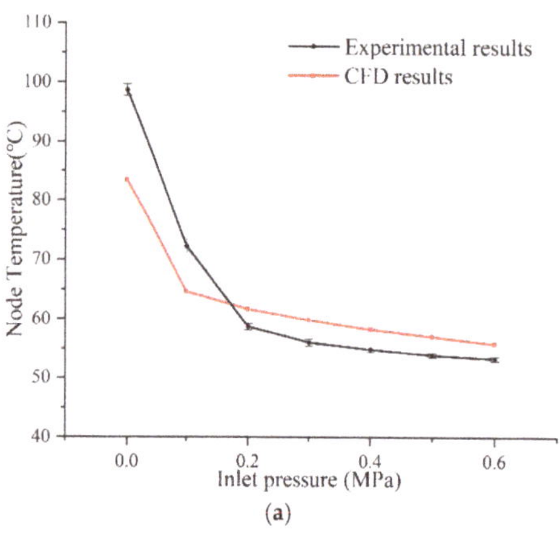
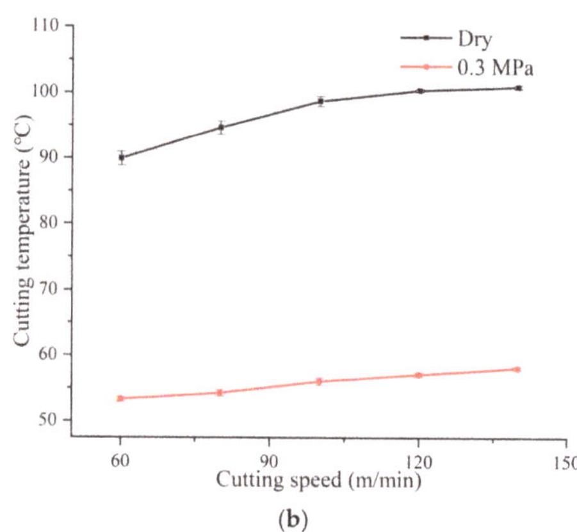

Figure 12. Effect of cooling conditions on cutting temperature. (**a**) cutting temperature under different spray pressure, (**b**) the influence of cutting speed on cutting temperature under dry cutting and internal spray cooling.

Figure 12b shows the cutting temperatures of the dry cutting and internal spray cooling at different cutting speeds. The experimental temperature gradually increases as the cutting speed increases. When the cutting speed are 60 m/min, 100 m/min and 140 m/min, the measured temperature of the internal spray cooling are 53.3 °C, 56.1 °C and 58.7°C, respectively, which are significantly lower than that of the dry cutting (89.9 °C, 98.7 °C and 100.9 °C). The cutting temperature under the internal spray cooling condition can be reduced by 41–44% compared to dry cutting with the cutting speed ranging from 60 m/min to 150 m/min.

4.3.2. Surface Roughness

Figure 13a shows the effect of the inlet pressure on the surface roughness of the workpiece internal surface at a cutting speed of 100 m/min. With the increase in the inlet pressure, the surface roughness of the workpiece decreases from 2.92 μm in dry cutting to 2.63 μm with a tool inlet pressure of 0.6 MPa. When the tool inlet pressure is lower than 0.2 MPa, the surface roughness decreases insignificantly with the increase in inlet pressure. This is mainly due to the low airflow velocity when the inlet pressure is small, which leads to the insignificant cooling and lubricating effect, and deteriorates the surface quality. However, the velocity of the air and coolant droplets increase when the tool inlet pressure is higher than 0.2 MPa, which improves the permeability of the coolant droplets and the chip removal, thereby enhancing the lubrication effect and convective heat transfer, and the surface quality as well.

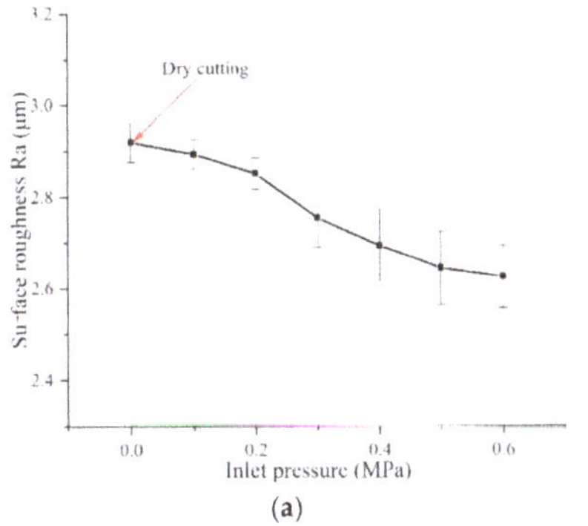
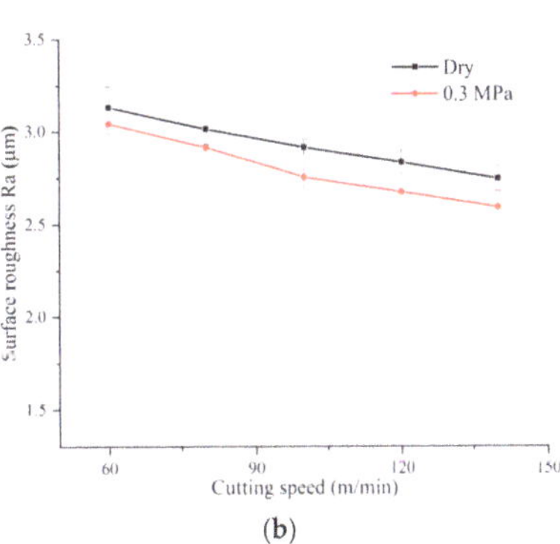

(a)   (b)

Figure 13. Effect of cooling conditions on machined surface roughness of workpiece. (**a**) Surface roughness of workpiece under different spray pressure at a cutting speed of 100 m/min, (**b**) The spray cooling and dry cutting roughness at different cutting speeds.

Figure 13b shows the effect of the cutting speed on the surface roughness of the workpiece internal surface. The surface roughness decreases with the increase in cutting speed ranging from 60 m/min to 140 m/min. Higher cutting speed leads to higher the cutting temperature, which subsequently softens the material and reduces the cutting force and vibration. Compared to the dry cutting, the internal spray cooling method could reduce the surface roughness of the workpiece by 0.1–0.25 μm.

4.3.3. Chip Morphology

Figure 14 shows the chip morphology under different cutting conditions. Overall, the chips produced under various cutting conditions can be characterised by the spiral feature. It can be seen from Figure 14a,c,g that the chip size and curling degree decreases with the increase in the cutting speed [30]. The chip length and curling degree are smaller under the spray cooling condition than that of the dry condition. Figure 14d–f shows the chips morphology under different tool inlet pressures at a cutting speed of 100 m/min. The chip length decreases gradually with the increase in tool inlet spray pressure due to the increasing fluid dynamic pressure applied on the chips, which made the chips more likely to be broken. For the internal turning, the small chips are beneficial to be removed by high-speed airflow to avoid chip blockage during turning.

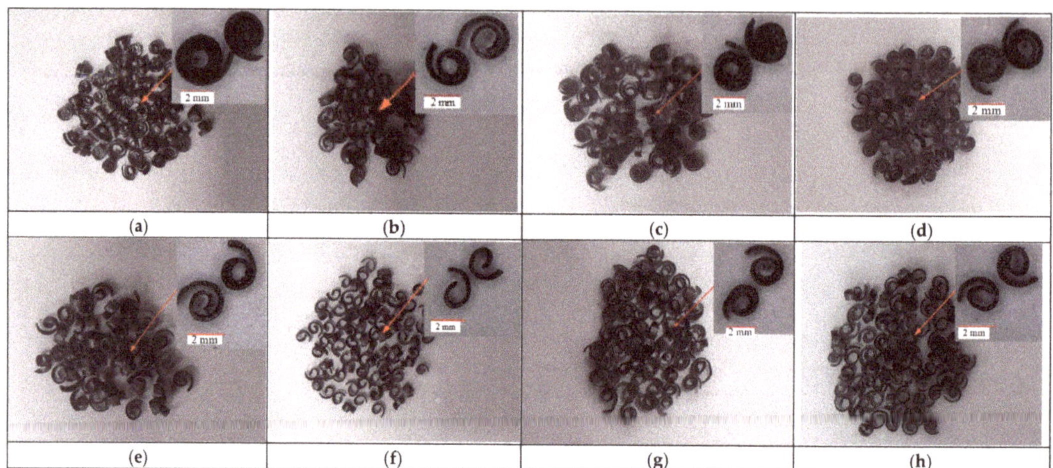

Figure 14. Chip morphology under different cutting conditions. (**a**) Vc = 60 m/min-Dry, (**b**) Vc = 60 m/min−0.3 MPa, (**c**) Vc = 100 m/min-Dry, (**d**) Vc = 100 m/min−0.1 MPa, (**e**) Vc = 100 m/min−0.3 MPa, (**f**) Vc = 100 m/min−0.6 MPa, (**g**) Vc = 140 m/min-Dry, (**h**) Vc = 140 m/min−0.3 MPa.

5. Conclusions

In order to improve the cooling efficiency of internal turning processes and achieve green cutting, an internal spray cooling turning tool was developed. The fluid-solid thermal coupling simulation model was established to simulate heat transfer during internal turning with internal spray cooling. The parameters of the cooling tool structure were optimized. The cutting experiments were conducted to investigate the cooling performance of the optimized internal spray cooling turning tool. The main conclusions can be drawn as follows:

- The structure parameters with the best cooling performance were determined, namely the diameter of upper nozzle is 3 mm, the diameter of lower nozzle is 1.5 mm, and the distance between upper nozzle and tool tip is 18.5 mm.
- With the increase in spray pressure, the velocity of air and droplets, and the convective heat transfer increase. The temperature decreases rapidly yet then mildly with the increasing inlet pressure due to the gradual saturation of the convective heat transfer. The internal turning experiments demonstrate that compared to the dry cutting, the cutting temperature with the internal spray cooling can be reduced by 41–44% with the inlet pressure of 0.3 MPa.
- The workpiece surface quality can be significantly improved under the spray cooling condition with larger tool inlet pressure. The internal spray cooling method can reduce the surface roughness of the workpiece by 0.1 µm–0.25 µm compared to that of the dry cutting condition. Particularly, the roughness of the inner surface of the workpiece can be reduced by 9.93% with the tool inlet pressure of 0.6 MPa.
- The increasing tool inlet pressure and cutting speed can efficiently decrease the chip length, and furthermore facilitate the chip removal to improve the cutting performance.

Author Contributions: Conceptualization, S.S. and L.L.; data curation, H.L. and X.C.; funding acquisition, S.S.; project administration, S.S. and L.L.; software, H.L.; writing—review and editing, H.L. and S.S. All authors have read and agreed to the published version of the manuscript.

Funding: This research was funded by the National Natural Science Foundation of China (Grant No. 52165058 and 51705153) and the Jiangxi Provincial Natural Science Foundation of China (Grant No. 20192BAB206029).

Institutional Review Board Statement: Not applicable.

Informed Consent Statement: Not applicable.

Data Availability Statement: Not applicable.

Conflicts of Interest: The authors declare no conflict of interest.

References

1. Suyama, D.I.; Diniz, A.E.; Pederiva, R. Tool vibration in internal turning of hardened steel using cBN tool. *Int. J. Adv. Manuf. Technol.* **2017**, *88*, 2485–2495. [CrossRef]
2. Basovich, S.; Arogeti, S. Identification and robust control for regenerative chatter in internal turning with simultaneous compensation of machining error. *Mech. Syst. Signal Process.* **2021**, *149*, 107208. [CrossRef]
3. Longbottom, J.M.; Lanham, J.D. Cutting temperature measurement while machining—A review. *Aircr. Eng. Aerosp. Technol.* **2005**, *77*, 122–130. [CrossRef]
4. Ezugwu, E.O.; Wang, Z.M. Titanium alloys and their machinability—A review. *J. Mater. Process. Technol.* **1997**, *68*, 262–274. [CrossRef]
5. Komanduri, R.; Hou, Z.B. Thermal modeling of the metal cutting process—Part II: Temperature rise distribution due to frictional heat source at the tool–chip interface. *Int. J. Mech. Sci.* **2001**, *43*, 57–88. [CrossRef]
6. Debnath, S.; Reddy, M.M.; Yi, Q.S. Environmental friendly cutting fluids and cooling techniques in machining: A review. *J. Clean. Prod.* **2014**, *83*, 33–47. [CrossRef]
7. Sharif, M.N.; Pervaiz, S.; Deiab, I. Potential of alternative lubrication strategies for metal cutting processes: A review. *Int. J. Adv. Manuf. Technol.* **2017**, *89*, 2447–2479. [CrossRef]
8. Shokrani, A.; Dhokia, V.; Newman, S.T. Energy conscious cryogenic machining of Ti-6Al-4V titanium alloy. *Proc. Inst. Mech. Eng. Part B J. Eng. Manuf.* **2018**, *232*, 1690–1706. [CrossRef]
9. Chiou, R.Y.; Lu, L.; Chen JS, J.; North, M.T. Investigation of dry machining with embedded heat pipe cooling by finite element analysis and experiments. *Int. J. Adv. Manuf. Technol.* **2007**, *31*, 905–914. [CrossRef]
10. Minton, T.; Ghani, S.; Sammler, F.; Bateman, R.; Fürstmann, P.; Roeder, M. Temperature of internally-cooled diamond-coated tools for dry-cutting titanium. *Int. J. Mach. Tools Manuf.* **2013**, *75*, 27–35. [CrossRef]
11. Meena, A.; Mansori, M.E. Study of dry and minimum quantity lubrication drilling of novel austempered ductile iron (ADI) for automotive applications. *Wear* **2011**, *271*, 2412–2416. [CrossRef]
12. Pereira, O.; Rodriguez, A.; Fernandez-Abia, A.I.; Barreiro, J.; Lopez de Lacalle, L.N. Cryogenic and minimum quantity lubrication for an eco-efficiency turning of AISI 304. *J. Clean. Prod.* **2016**, *139*, 440–449. [CrossRef]
13. Li, B.D.; Rui, Z.Y. Friction and wear behaviours of YG8 sliding against austempered ductile iron under dry, chilled air and minimal quantity lubrication conditions. *Adv. Mech. Eng.* **2019**, *11*, 1687814019847064. [CrossRef]
14. Das, A.; Padhan, S.; Das, S.R.; Alsoufi, M.S.; Ibrahim AM, M.; Elsheikh, A. Performance assessment and chip morphology evaluation of austenitic stainless steel under sustainable machining conditions. *Metals* **2021**, *11*, 1931. [CrossRef]
15. Zeilmann, R.P.; Weingaertner, W.L. Analysis of temperature during drilling of Ti6Al4V with minimal quantity of lubricant. *J. Mater. Process. Technol.* **2006**, *179*, 124–127. [CrossRef]
16. Jessy, K.; Dinakaran, D.; Seshagiri Rao, V. Influence of different cooling methods on drill temperature in drilling GFRP. *Int. J. Adv. Manuf. Technol.* **2015**, *76*, 609–621. [CrossRef]
17. Li, Y.; Wu, W. Investigation of drilling machinability of compacted graphite iron under dry and minimum quantity lubrication (MQL). *Metals* **2019**, *9*, 1095. [CrossRef]
18. Qin, X.D.; Liu, W.T.; Li, S.T.; Tong, W.; Ji, X.L.; Meng, F.J.; Liu, J.H.; Zhao, E.H. A comparative study between internal spray cooling and conventional external cooling in drilling of Inconel 718. *Int. J. Adv. Manuf. Technol.* **2019**, *104*, 4581–4592. [CrossRef]
19. Shu, S.R.; Zhang, Y.; He, Y.Y.; Zhang, H. Design of a novel turning tool cooled by combining circulating internal cooling with spray cooling for green cutting. *J. Adv. Mech. Des. Syst. Manuf.* **2021**, *15*, JAMDSM0003. [CrossRef]
20. Gan, Y.Q.; Wang, Y.Q.; Liu, K.; Wang, S.Q.; Yu, Q.B.; Che, C.; Liu, H.B. The development and experimental research of a cryogenic internal cooling turning tool. *J. Clean. Prod.* **2021**, *319*, 128787. [CrossRef]
21. Obikawa, T.; Asano, Y.; Kamata, Y. Computer fluid dynamics analysis for efficient spraying of oil mist in finish-turning of Inconel 718. *Int. J. Mach. Tools Manuf.* **2009**, *49*, 971–978. [CrossRef]
22. Duchosal, A.; Serra, R.; Leroy, R. Numerical study of the inner canalization geometry optimization in a milling tool used in micro quantity lubrication. *Mech. Ind.* **2014**, *15*, 435–442. [CrossRef]
23. Zhang, C.L.; Zhang, S.; Yan, X.F.; Zhang, Q. Effects of internal cooling channel structures on cutting forces and tool life in side milling of H13 steel under cryogenic minimum quantity lubrication condition. *Int. J. Adv. Manuf. Technol.* **2016**, *83*, 975–984. [CrossRef]
24. Peng, R.T.; Jiang, H.J.; Tang, X.Z.; Huang, X.F.; Xu, Y.; Hu, Y.B. Design and performance of an internal-cooling turning tool with micro-channel structures. *J. Manuf. Process.* **2019**, *45*, 690–701. [CrossRef]
25. Zaman, P.B.; Dhar, N.R. Design and evaluation of an embedded double jet nozzle for MQL delivery intending machinability improvement in turning operation. *J. Manuf. Process.* **2019**, *44*, 179–196. [CrossRef]

26. Kumar, A.; Singh, G.; Aggarwal, V. Analysis and optimization of nozzle distance during turning of EN-31 steel using minimum quantity lubrication. *Mater. Today Proc.* **2021**, *49*, 1360–1366. [CrossRef]
27. Liu, J.Y.; Han, R.D.; Zhang, L. Study on the effect of jet flow parameters with water vapor as coolants and lubricants in green cutting. *Ind. Lubr. Tribol.* **2007**, *59*, 278–284. [CrossRef]
28. Shih, T.H.; Liou, W.W.; Shabbir, A. A new k-ϵ eddy viscosity model for high reynolds number turbulent flows. *Comput. Fluids* **1995**, *24*, 227–238. [CrossRef]
29. Dhar, N.R.; Kamruzzaman, M.; Ahmed, M. Effect of minimum quantity lubrication (MQL) on tool wear and surface roughness in turning AISI-4340 steel. *J. Mater. Process. Technol.* **2006**, *172*, 299–304. [CrossRef]
30. Yang, H.; Wang, R.; Guo, Z.; Lin, R.; Wei, S.; Weng, J. Cutting performance of multicomponent AlTiZrN-coated cemented Carbide (YG8) tools during milling of high-chromium cast iron. *Coatings* **2022**, *12*, 686. [CrossRef]

Article

Multisensor Feature Fusion Based Rolling Bearing Fault Diagnosis Method

Jinyu Tong [1,2], Cang Liu [2], Haiyang Pan [2] and Jinde Zheng [2,*]

1 Anhui Province Engineering Laboratory of Intelligent Demolition Equipment,
   Anhui University of Technology, Ma'anshan 243002, China; pantc2006@163.com
2 School of Mechanical Engineering, Anhui University of Technology, Ma'anshan 243002, China;
   liucang0929@163.com (C.L.); pansea@sina.cn (H.P.)
* Correspondence: lqdlzheng@126.com

Abstract: To fully utilize the fault information and improve the diagnosis accuracy of rolling bearings, a multisensor feature fusion method is proposed. The method contains two steps. First, the intrinsic mode function (IMF) of each sensor vibration signal is calculated by variational mode decomposition (VMD), and the redundant information such as noise is eliminated. Then, the time-domain, frequency-domain and multiscale entropy features are extracted based on the preferred IMF and fused into one multidomain feature dataset. In the second step, the deep autoencoder network (DAEN) is constructed and the multisensor fusion features of the first step are used as input of the DAEN, and the multisensor fusion features are further extracted and classified. The experimental results show that the proposed model has a higher classification accuracy compared with the existing methods.

Keywords: fault diagnosis; autoencoder network; multisensor; feature fusion; rolling bearing

Citation: Tong, J.; Liu, C.; Pan, H.; Zheng, J. Multisensor Feature Fusion Based Rolling Bearing Fault Diagnosis Method. *Coatings* **2022**, *12*, 866. https://doi.org/10.3390/coatings12060866

Academic Editor: María Dolores Fernández Ramos

Received: 20 May 2022
Accepted: 16 June 2022
Published: 19 June 2022

Publisher's Note: MDPI stays neutral with regard to jurisdictional claims in published maps and institutional affiliations.

1. Introduction

As a critical component of rotating machine, rolling bearings have the advantages of high efficiency, low friction resistance and convenient assembly. Furthermore, their performance directly affects the operation of all the equipment. Therefore, knowing how to fully exploit the fault features from the complex vibration signals and carry out pattern recognition is of great significance [1,2].

The mainstream methods of fault diagnosis only focus on the application of a single sensor [3–5]. The commonly used sensor is the vibration acceleration sensor, which can measure the relationship between the vibration amplitude and time. However, more and more studies have shown that, for a complex mechanical system, the fault information contained in a single sensor is limited, and accurate condition monitoring and fault diagnosis cannot be performed [6–8]. The application of multiple sensor technologies in fault diagnosis makes it possible to study fault diagnosis based on multiple sensors. Wang et al. [9] proposed a mixture of Gaussians and variational auto-encoders (Mix-VAEs) fault diagnosis method, which can fully utilize the redundancy and complementarity of multisensor information. Chen et al. [10] proposed an stack auto-encoder and deep belief network (SAE-DBN) based multisensor fusion method, and verified the effectiveness through a bearing fault experiment. Shi et al. [11] proposed a two-stage multisensor fusion method to achieve accurate diagnosis of hydraulic directional valve faults. The above studies show that compared with a single sensor, multisensor information fusion technology can further improve the accuracy and reliability of diagnosis.

Multisensor information fusion technology includes data-level fusion, feature-level fusion and decision-level fusion, which have their own advantages and limitations [12–14]. The advantage of data-level fusion is that the raw signals of multiple sensors can be directly fused. Unfortunately, the raw data usually contains a lot of redundant information, and the data-level fusion method cannot take full advantage of the complementarity between the information of multiple sensors. Furthermore, the interpretability of the data is poor. Jing et al. [15] directly fused data from multisensors to construct a deep network for planetary gearbox fault diagnosis. Huang et al. [16] proposed a multisensor data fusion method to solve the problem of multisource remote sensing data fusion.

In order to make up for the deficiencies of data-level fusion methods and eliminate redundant information from multiple sensors, data-level fusion methods can be combined with feature extraction methods. First, the data from each sensor is transformed into a high-dimensional feature representation, and then, fusion is performed at the feature level, and this fusion method is called feature-level fusion [17,18]. Li et al. [19] proposed a fault diagnosis method based on a feature fusion covariance matrix and Riemann kernel ridge regression. Wang et al. [20] proposed a multisource sensor feature fusion method based on a convolutional neural network for mechanical fault diagnosis. Jiang et al. [21] extracted various entropy values of vibration signals using information entropy theory, and established a feature-level fusion model to classify faults. One of the advantages of feature fusion is that it can flexibly choose where to fuse, but it cannot eliminate the effect of high correlations between different sensor features.

In decision-level fusion, the basic learning model is first trained with different sensor signals, and then the output results of multiple models are fused through decision strategies. The errors of fusion models come from different basic learning models, which are often ir-relevant and do not affect each other, and will not cause further accumulation of errors. Therefore, the decision-level fusion method is favored. Common decision fusion methods [22,23] include the voting method and D-S evidence theory. Li et al. [24] proposed an enhanced weighted voting combination strategy with specific category threshold to realize multisensor decision fusion. Basir et al. [25] constructed a multisensor-based model according to D-S evidence theory to solve the problem of engine fault diagnosis. Zhao et al. [26] proposed a new distributed distance measurement method to measure the conflict between evidence based on an improved evidence theory algorithm. The decision-level fusion method is very sensitive to the selection of voting fusion rules, which directly determines the fusion result.

For the fault diagnosis of multisensor fusion, a unified and effective fusion model and algorithm has not yet been established, and various proposed models are still in the exploratory stage. From the above discussion, it can be seen that feature-level fusion is more flexible and convenient, not only to select information that can characterize fault features, but also to fuse at multiple locations. Furthermore, deep learning has the ability to learn features directly from raw signals, which largely overcomes the loss of effective information in feature-level fusion. Therefore, this paper proposes a multisensor feature fusion method combined with feature-level fusion and the deep learning method, and applies them to the fault diagnosis of rolling bearings under different working conditions. The proposed feature fusion method provides a more effective means for the deep mining of fault signals. The main contributions of this paper are as follows:

(1) A multisensor signals-based feature fusion method is proposed for one-dimensional vibration signals.
(2) The vibration signal of each sensor is preprocessed with VMD, and the time domain, frequency domain and multiscale entropy features of the signal are extracted and fused into one multidomain feature dataset.
(3) To promote further fusion of features, a novel deep autoencoder network is proposed for feature extraction and classification.

The rest of the paper is organized as follows. Section 2 reviews the AE. In Section 3, the proposed model is described in detail. Section 4 gives a detailed analysis and discussion of the experimental diagnosis results of rolling bearings. Section 5 presents the conclusions and possible future research directions.

2. Theoretical Basis

Autoencoder

Autoencoders (AE) can minimize the reconstruction error of input and output and are unsupervised neural networks. The structure of AE is shown in Figure 1. It consists of an input layer, a hidden layer and an output layer. The input layer and the hidden layer constitute the encoder, and the hidden layer and the output layer constitute the decoder. The encoder converts the high-dimensional input data into a low-dimensional feature representation, and the decoder converts the feature representation into a reconstructed form of the input data.

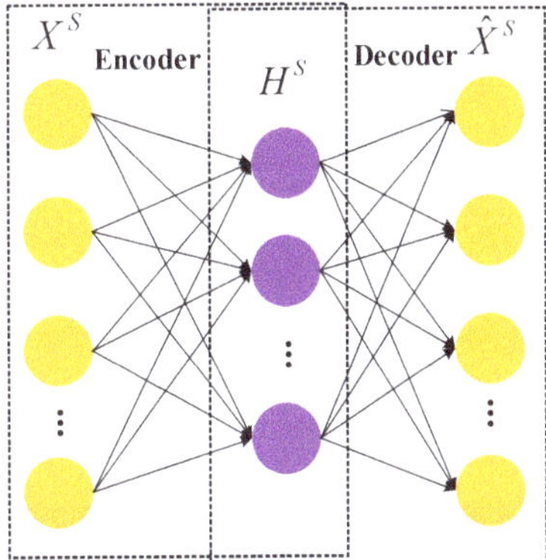

Figure 1. Structure of AE.

The encoder maps raw input signal **X** to the hidden layer feature **H**. The process is as follows:

$$\mathbf{H} = r_f(\mathbf{WX} + \mathbf{b})\tag{1}$$

The decoder reconstructs the hidden layer feature **H** to obtain the output vector **X̂**. The process is as follows:

$$\hat{\mathbf{X}} = r_g(\mathbf{W'X} + \mathbf{b'})\tag{2}$$

where **W** and **W'** are the weight matrix, **b** and **b'** are the bias matrix r_f and r_g are the activation function.

The reconstruction error of AE is:

$$L(\mathbf{X}, \hat{\mathbf{X}}) = \frac{1}{2}\|\mathbf{X} - \hat{\mathbf{X}}\|^2\tag{3}$$

where $\|\bullet\|$ represents the norm.

Therefore, the total loss function for S sample is:

$$J(\mathbf{W}, \mathbf{b}) = \frac{1}{S}\sum_{n=1}^{S} L(\mathbf{X}, \hat{\mathbf{X}})\tag{4}$$

3. Proposed Method

In this section, a feature fusion model based on multisensor signals is proposed and applied to rolling bearing fault diagnosis.

3.1. Fusion Model Architecture for Multisensor Signals

The proposed method consists of two steps. The first step is multisensor feature fusion, where the IMF of each sensor vibration signal is calculated by VMD [27]. Then, time-domain, frequency-domain and multiscale entropy features are extracted based on the preferred IMF and fused into a multidomain feature dataset. In the second step, the DAEN is constructed and the multisensor fusion features of the first step are used as inputs of the DAEN. Then, the multisensor fusion features are further extracted and classified.

3.2. Implementation Process

3.2.1. Multisensor Feature Fusion

The proposed feature fusion method is as follows:

(1) The vibration signal $\mathbf{X}^1_{l\times1}, \mathbf{X}^2_{l\times1}, \cdots, \mathbf{X}_{l\times1}{}^n$ is collected from n sensors of different directions, where l is the sample length.

(2) Take the data length i as a sample and divide $\mathbf{X}^1_{l\times1}, \mathbf{X}^2_{l\times1}, \cdots, \mathbf{X}_{l\times1}{}^n$ into $\mathbf{X}^1_{m\times i}, \mathbf{X}^2_{m\times i}, \cdots, \mathbf{X}^n_{m\times i}$, where m is the number of samples.

(3) Using the VMD to decompose $\mathbf{X}^1_{m\times i}, \mathbf{X}^2_{m\times i}, \cdots, \mathbf{X}^n_{m\times i}$, a number of IMF components of each sensor are obtained, and base on the decomposition results, the first few components already contain the main information of the raw signal [28], so in this paper, we take the modal number $k = 3$ and decompose it to obtain $\mathbf{X}^1_{m\times3\times1024}, \mathbf{X}^2_{m\times3\times1024}, \cdots, \mathbf{X}^n_{m\times3\times1024}$.

(4) Feature extraction is performed for IMF components, and 12 time-domain features and five frequency-domain features [29] are extracted for each IMF component. To further reflect the degree of self-similarity and complexity of vibration signals under different scale factors of the same time series, five multiscale entropy values are extracted for each IMF component, denoted as $\mathbf{X}^1_{m\times3\times22}, \mathbf{X}^2_{m\times3\times22}, \cdots, \mathbf{X}^n_{m\times3\times22}$.

(5) The raw feature multidomain set is formed by fusing the proposed features, denoted as $\mathbf{X}^1_{m\times66}, \mathbf{X}^2_{m\times66}, \cdots, \mathbf{X}^n_{m\times66}$, and further fusing the raw feature multidomain set of sensors in each direction to obtain $\widetilde{\mathbf{X}} = \left[\mathbf{X}^1_{m\times66}, \mathbf{X}^2_{m\times66}, \cdots, \mathbf{X}^n_{m\times66}\right], \widetilde{\mathbf{X}} \in \mathbb{R}_{m\times66\times n}$.

3.2.2. Deep Feature Learning and Classification

To enhance the performance of multisensor feature fusion, the DAEN model is proposed for deep feature learning and classification in this section. The proposed DAEN model is a multilayer neural network, which is composed of multiple stacked AE and a Softmax classification layer. The structure of DAEN is shown in Figure 2.

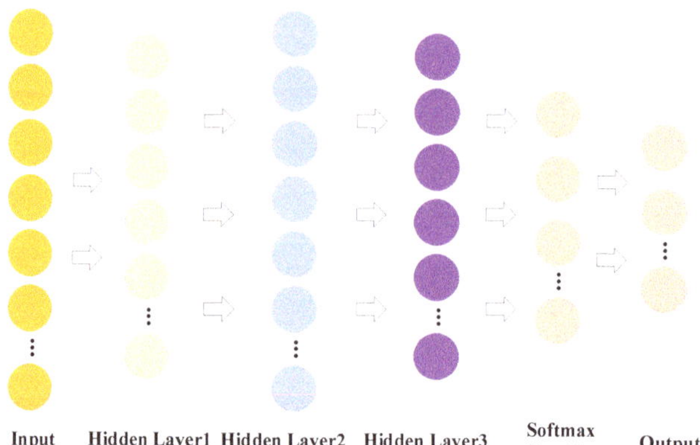

Figure 2. Structure of the proposed DAEN.

DAEN uses the Sigmoid activation function for nonlinear mapping [30]. The Sigmoid activation function is defined as follows:

$$\mathrm{Sigmoid}(x) = \frac{1}{1+e^{-x}} \tag{5}$$

The output of DAEN hidden layer is:

$$h_i = \cfrac{1}{1 + e^{-(\sum\limits_{j=1}^{N} w_{ij}x_j + b_j)}} \tag{6}$$

where w_{ij} is the connection weight between node i at layer L and node j at layer $L + 1$, and b_j is the bias of the hidden layer node j.

The most commonly used loss function of AE is the mean square error [31], which is defined as:

$$L(x, \hat{x}) = \sum_{i=1}^{S} (x - \hat{x})^2 \tag{7}$$

Then the loss function of the proposed DAEN model can be expressed as:

$$J(w, b) = \sum_{t=1}^{S} \left(x^i - \hat{x}^i\right)^2 + rR(w, b) \tag{8}$$

where the first term is the mean square error loss, the second term is the penalty term and r is the sparse penalty factor.

The training process of DAEN consists of unsupervised training and fine-tuning. The process is as follows:

(1) The first stage fused feature $\widetilde{\mathbf{X}}$ is used as the input of the DAEN;
(2) Forward propagation. The hidden layer features of the first AE h_1 is used as the input of the second AE for unsupervised training until all hidden layers are trained;
(3) The backpropagation (BP) algorithm [32] is used for supervised fine-tuning to further optimize all the weights and biases;
(4) The last hidden layer feature, h_n, of the DAEN is fed into the Softmax classifier;
(5) The classification result is obtained.

3.3. Rolling Bearing Fault Diagnosis Process Based on the Proposed Method

Based on the proposed method, the process of the rolling bearing fault diagnosis method is as follows:

(1) Acquisition of rolling bearing vibration data from multiple sensors;
(2) The vibration signal of each sensor is preprocessed with VMD, and the 22 features of the signal are extracted based on the preferred IMF;
(3) The extracted feature is fused into multidomain feature dataset;
(4) The multidomain feature dataset is divided into either a training dataset or a testing dataset, according to the set ratio;
(5) The DAEN model is constructed. The parameters of the DAEN model are initialized, the training dataset is taken as the input to the model and the model loss function is minimized;
(6) The test dataset is fed into the trained DAEN model to obtain the test accuracy.

4. Experiment

4.1. Rolling Bearing Test Bench

To verify the superiority of the proposed method, the experimental data are obtained from the self-made rolling bearing fault test bench belonging to Anhui University of Technology, as shown in Figure 3. The experimental bearing is 6206-2RS1 SKF. Different depth faults are manufactured on the inner ring, outer ring and rolling ball for the rolling bearing by electric sparkline cutting technology. Figure 4 presents four different health states for rolling bearing.

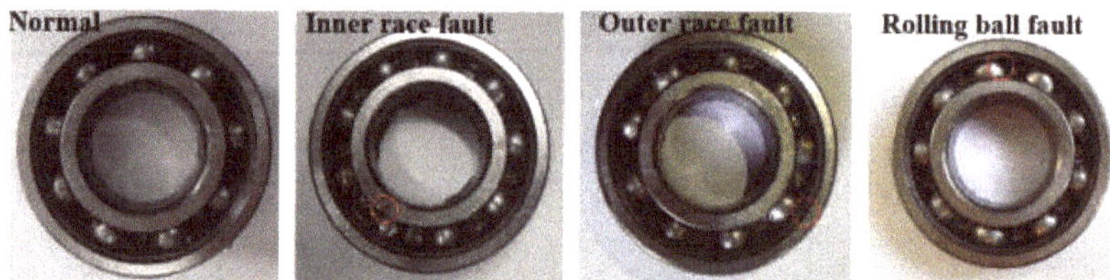

Figure 3. Schematic diagram of rolling bearing test rig.

Figure 4. Different health states of rolling bearings.

4.2. Rolling Bearing Multisensor Signals

The sampling frequency is set to 8192 Hz. When the load is 5 KN and the motor speed is 300 r/min, the signals of the rolling bearings in different health states are collected. Figure 5 shows the time-domain vibration signals of rolling bearings from three different directional sensors. The signals collected from each directional sensor contain six health states, including two types of inner ring faults with the fault depth of 0.3 and 0.4 mm, two types of outer ring faults with the fault depth of 0.2 and 0.3 mm, and one type of rolling bearing normal state.

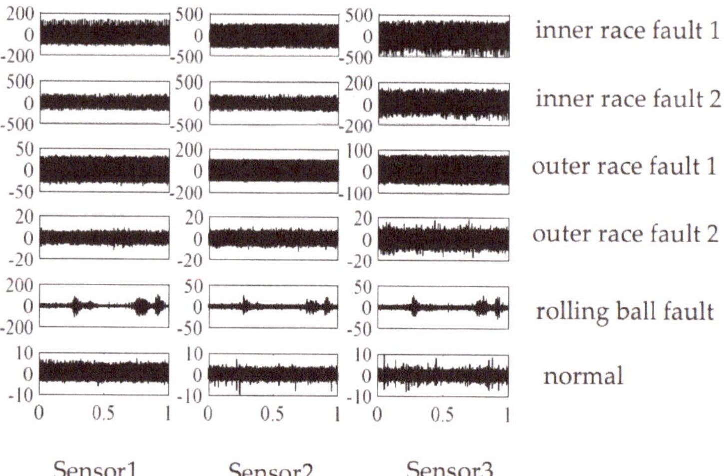

Figure 5. Time-domain vibration signals of rolling bearing from three different sensors.

4.3. Dataset Construction

Under the same fault type, 1024 data points are taken as one sample, and 100 samples are taken for each fault type randomly, which comes to 600 samples in total in this experiment. Three IMF components were obtained by decomposing each sample with VMD, and 17 time-domain and frequency-domain features were extracted for each IMF component, as well as five multiscale entropy values. After feature extraction, a sample of 1024 data points is changed into a sample of 66 data points as the input of the proposed model and the comparison model. 90% of them are randomly divided into the training set and 10% into the testing set, as shown in Table 1. That is, each category obtained a training sample of faults with a data dimension of 90 × 66 and a test sample of 10 × 66. After fusion at the one-dimensional feature level, each type of fault of the multisensor signal obtained a training sample of 90 × 198 and a test sample of 10 × 198.

Table 1. Bearing dataset information.

Fault Type	Fault Depth/mm	Size of Training Dataset	Size of Testing Dataset	Label
Inner race fault 1	0.3	90	10	1
Inner race fault 2	0.4	90	10	2
Outer race fault 1	0.2	90	10	3
Outer race fault 2	0.3	90	10	4
Rolling ball fault	0.2	90	10	5
Normal	0	90	10	6

4.4. Comparative Experiments and Analysis of Results

4.4.1. The Feasibility and Effectiveness of Multisensor Collaborative Diagnosis

In order to prove the feasibility and effectiveness of multisensor collaborative diagnosis, the vibration signals in three directions of sensor and multisensor fusion signals are input into DAEN for comparison according to the dataset construction method in Section 4.3. Through many experiments, the structure of DAEN based on a single sensor signal is set as [66 50 30 10 6], and the structure of DAEN based on multisensor signals is set as [198 50 30 10 6], i.e., one input layer, three hidden layers and one output layer [30]. The initial learning rate of DAEN is 0.01, the maximum number of iterations is 100, the sparse parameter r is 0.01 and the sparse penalty coefficient is 0.13. In order to eliminate the influence of random errors, 10 experiments were conducted for each method, and the mean and standard deviation of the 10 experimental results were used as the evaluation index of the method. A total of 10 experimental results were compared, as shown in Figure 6, and the mean accuracy and standard deviation of the 10 experiments are shown in Table 2.

As can be seen from Table 2, compared with single sensor 1~sensor 3, the diagnosis accuracy based on multisensor fusion is improved by 4.43%, 10.10% and 6.27%, respectively. The above results show that the diagnosis effect based on multisensor fusion signal is significantly better than that of the single sensor fusion signal, which proves that multisensor signal co-operative diagnosis is feasible and effective. At the same time, we can see from Table 2 that the diagnostic accuracy of different sensors is very different, indicating that the fault information contained in different sensor signals is different. When different sensors co-operate in a diagnosis, more accurate and reliable results can be provided.

4.4.2. Verification of the Superiority of the Proposed Method

To verify the performance of the proposed model, we compared stacked sparse autoencoder (SSAE), traditional machine learning method random forest (RF) and support vector machine (SVM). For fair comparison, the network structure of SSAE is the same as the proposed method, and the sparse parameter in SSAE is set to 0.2 and the sparse penalty coefficient is set to 0.15. The maximum depth of RF is set to 2, which contains 200 trees. The kernel function of SVM adopts RBF function. The penalty factor and kernel function parameters are set to 10 and 0.01, respectively.

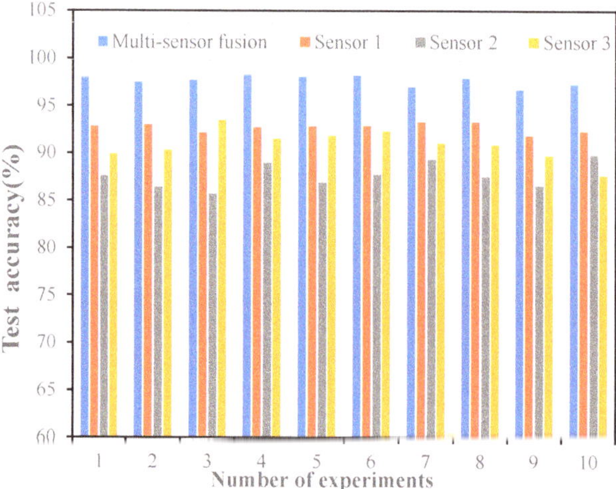

Figure 6. Comparison of 10 experiment results for different sensors' datasets.

Table 2. The diagnostic result of different sensor dataset.

Method	Average Test Accuracy (%)	Standard Deviation
Multisensor fusion (The proposed method)	97.55	0.485
Senor 1	93.12	0.589
Senor 2	87.45	1.418
Senor 3	91.28	1.803

In order to eliminate the influence of random errors, 10 experiments were conducted for each method, and the mean and standard deviation of the 10 experimental results were used as the evaluation index of the method. 10 experimental results were compared, as shown in Figure 7, and the mean accuracy and standard deviation of the 10 experiments are shown in Table 3.

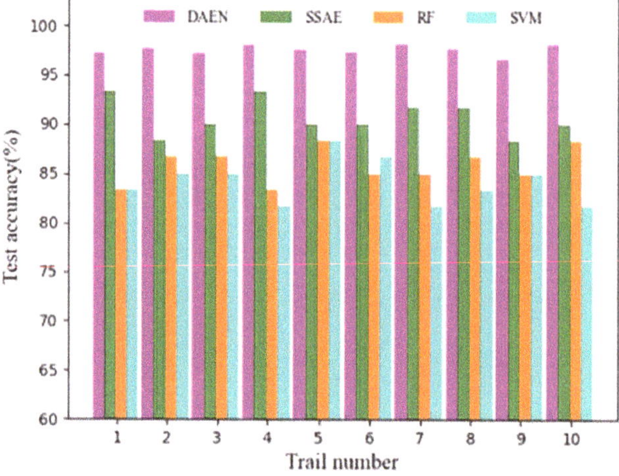

Figure 7. Comparison of experimental results.

Table 3. The average accuracy and standard deviation.

Method	Average Test Accuracy(%)	Standard Deviation
The proposed method	97.55	0.485
SSAE	90.67	1.792
RF	85.83	1.801
SVM	84.16	2.255

As can be seen from Figure 7 and Table 3, among the four methods, as traditional machine learning methods, the diagnostic results of RF and SVM in 10 experiments are lower than the other two autoencoder networks. This shows that the traditional machine learning method has a weak feature extraction ability and a low generalization ability when dealing with complex signals, and it is difficult to obtain a good diagnosis effect. Among the two autoencoder networks, the diagnostic accuracy of SSAE is 6.88% lower than that of DAEN, and the standard deviation is increased by 71.26%, which indicates that SSAE has a weaker feature extraction ability. The proposed method has the highest diagnostic accuracy and the lowest standard deviation in 10 experiments, indicating that the proposed method can mine fault-sensitive features effectively, make more, full use of multisensor information and improve the diagnostic effect and stability.

Figure 8 shows the confusion matrix of the first trial of the proposed method. The horizontal co-ordinates of the confusion matrix plot are the true labels, the vertical co-ordinates are the predicted labels and the numbers on the diagonal lines indicate the classification accuracy of the proposed method for each type of sample. From Figure 8, it can be seen that the proposed method can identify 100% of the five conditions of inner ring fault 2, outer ring fault 1, outer ring fault 2, rolling ball failure and normal condition for the rolling bearing dataset of six health conditions. The only misclassification occurred in the inner ring fault 1 sample.

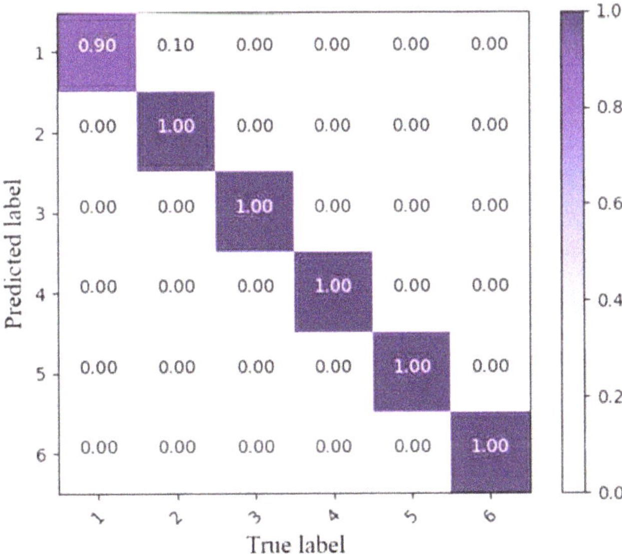

Figure 8. Confusion matrix of the first trial of the proposed method.

The t-distribution neighborhood embedding (t-SNE) algorithm [33] is adopted for feature visualization. T-SNE method is used to draw scatter plots of the raw data, respectively. The output of the features from the Softmax layer of the proposed method is shown in Figure 9. From Figure 9, it can be seen that the raw time-domain signal contains too much redundant information, and the features of all categories are difficult to distinguish. In contrast, the features extracted by the proposed method in the Softmax layer are easier to distinguish and show a better classification effect, i.e., the same fault features are clustered according to the same center and different fault features are distinguished, which proves the better performance of the proposed method.

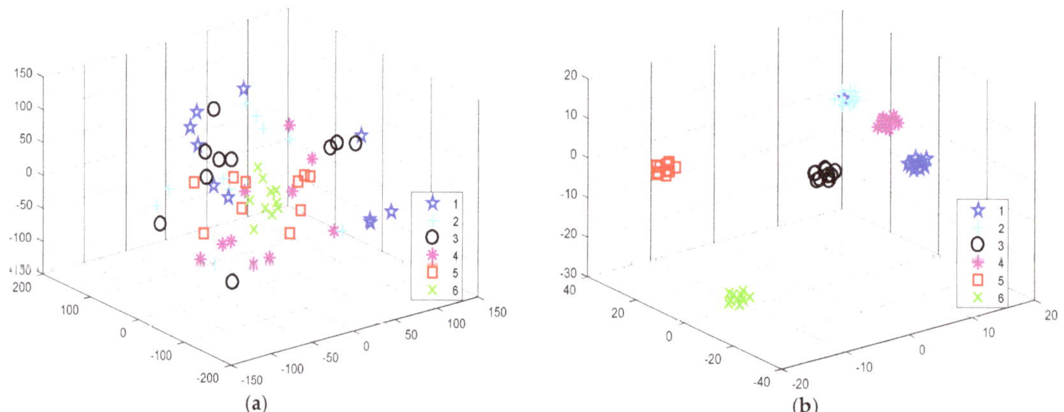

Figure 9. Feature visualization. (**a**) Feature visualization of raw signal; (**b**) Feature visualization of Softmax layer.

5. Conclusions

In order to improve the fault diagnosis accuracy of rolling bearings, a novel multisensor feature fusion method is proposed in this paper. VMD is used to decompose multiple sensor signals, which reduces the redundant information contained in the raw signals. The multidomain features of each single sensor are fused at the feature-level, and the complementary information among multiple sensors is effectively utilized. The depth features of multisensor are further learned and fused with the constructed DAEN. The diagnosis effect of the proposed method is better than that of a single sensor, showing better robustness and providing a more effective means for fault signal deep mining and multisensor information fusion.

Author Contributions: Conceptualization, J.T.; methodology, J.T.; software, C.L.; validation, C.L.; formal analysis, H.P.; investigation, J.T. and C.L.; resources, H.P.; data curation, H.P.; writing—original draft preparation, C.L.; writing—review and editing, J.T.; visualization, H.P.; supervision, J.Z.; project administration, J.Z.; funding acquisition, J.Z. All authors have read and agreed to the published version of the manuscript.

Funding: This work was supported by the Open Project of Anhui Province Engineering Laboratory of Intelligent Demolition Equipment (No. APELIDE2021B006), the National Natural Science Foundation of China (No. 51975004).

Institutional Review Board Statement: Not applicable.

Informed Consent Statement: Not applicable.

Data Availability Statement: Not applicable.

Conflicts of Interest: The authors declare no conflict of interest.

References

1. Wang, X.; Shen, C.; Xia, M.; Wang, D.; Zhu, J.; Zhu, Z. Multi-scale deep intra-class transfer learning for bearing fault diagnosis. *Reliab. Eng. Syst. Saf.* **2020**, *202*, 107050. [CrossRef]
2. Al-Bugharbee, H.; Trendafilova, I. A fault diagnosis methodology for rolling element bearings based on advanced signal pretreatment and autoregressive modeling. *J. Sound Vib.* **2016**, *369*, 246–265. [CrossRef]
3. Zhang, W.; Li, C.; Peng, G.; Chen, Y.; Zhang, Z. A deep convolutional neural network with new training methods for bearing fault diagnosis under noisy environment and different working load. *Mech. Syst. Signal Processing* **2017**, *100*, 439–453. [CrossRef]
4. Hu, Z.; Han, T.; Bian, J.; Wang, Z.; Chen, L.; Zhang, W.; Kong, X. A deep feature extraction approach for bearing fault diagnosis based on multi-scale convolutional autoencoder and generative adversarial networks. *Meas. Sci. Technol.* **2022**, *33*, 065013. [CrossRef]
5. Shao, H.; Jiang, H.; Li, X.; Wu, S. Intelligent fault diagnosis of rolling bearing using deep wavelet auto-encoder with extreme learning machine. *Knowl. Based Syst.* **2018**, *140*, 1–14.
6. Shao, H.; Lin, J.; Zhang, L.; Galar, D.; Kumar, U. A novel approach of multisensory fusion to collaborative fault diagnosis in maintenance. *Inf. Fusion* **2021**, *74*, 65–76. [CrossRef]
7. Shan, P.; Lv, H.; Yu, L.; Ge, H.; Li, Y.; Gu, L. A multisensor data fusion method for ball screw fault diagnosis based on convolutional neural network with selected channels. *IEEE Sens. J.* **2020**, *20*, 7896–7905. [CrossRef]
8. Bigdeli, B.; Pahlavani, P.; Amirkolaee, H.A. An ensemble deep learning method as data fusion system for remote sensing multisensor classification. *Appl. Soft Comput.* **2021**, *110*, 107563. [CrossRef]
9. Wang, C.; Xin, C.; Xu, Z.; Qin, M.; He, M. Mix-VAEs: A novel multisensor information fusion model for intelligent fault diagnosis. *Neurocomputing* **2022**, *492*, 234–244. [CrossRef]
10. Chen, Z.; Li, W. Multisensor feature fusion for bearing fault diagnosis using sparse autoencoder and deep belief network. *IEEE Trans. Instrum. Meas.* **2017**, *66*, 1693–1702. [CrossRef]
11. Shi, J.; Yi, J.; Ren, Y.; Li, Y.; Zhong, Q.; Tang, H.; Chen, L. Fault diagnosis in a hydraulic directional valve using a two-stage multi-sensor information fusion. *Measurement* **2021**, *179*, 109460. [CrossRef]
12. Wang, X.; Lu, S.; Chen, K.; Wang, Q.; Zhang, S. Bearing fault diagnosis of switched reluctance motor in electric vehicle powertrain via multi-sensor data fusion. *IEEE Trans. Ind. Inform.* **2022**, *18*, 2452–2464. [CrossRef]
13. Yu, J. Mechatronics fault prediction and diagnosis based on multi sensor information fusion. *J. Phys. Conf. Ser.* **2021**, *1982*, 012100. [CrossRef]
14. Wang, D.; Li, Y.; Jia, L.; Song, Y.; Liu, Y. Novel three-stage feature fusion method of multimodal data for bearing fault diagnosis. *IEEE Trans. Instrum. Meas.* **2021**, *70*, 1–10. [CrossRef]
15. Jing, L.; Wang, T.; Zhao, M.; Wang, P. An adaptive multi-sensor data fusion method based on deep convolutional neural networks for fault diagnosis of planetary gearbox. *Sensors* **2017**, *17*, 414. [CrossRef]
16. Huang, M.; Liu, Z.; Tao, Y. Mechanical fault diagnosis and prediction in IoT based on multi-source sensing data fusion. *Simul. Model. Pract. Theory* **2019**, *102*, 101981. [CrossRef]
17. Zhang, Y.; Li, C.; Wang, R.; Qian, J. A novel fault diagnosis method based on multi-level information fusion and hierarchical adaptive convolutional neural networks for centrifugal blowers. *Measurement* **2021**, *185*, 109970. [CrossRef]
18. Shao, H.; Jiang, H.; Wang, F.; Zhao, H. An enhancement deep feature fusion method for rotating machinery fault diagnosis. *Knowl.-Based Syst.* **2017**, *119*, 200–220. [CrossRef]
19. Li, X.; Zhong, X.; Shao, H.; Han, T.; Shen, C. Multi-sensor gearbox fault diagnosis by using feature-fusion covariance matrix and multi-Riemannian kernel ridge regression. *Reliab. Eng. Syst. Saf.* **2021**, *216*, 108018. [CrossRef]
20. Wang, N.; Li, H.; Wu, F.; Zhang, R.; Gao, F. Fault diagnosis of complex chemical processes using feature fusion of a convolutional network. *Ind. Eng. Chem. Res.* **2021**, *60*, 2232–2248. [CrossRef]
21. Jiang, Q.; Shen, Y.; Li, H.; Xu, F. New fault recognition method for rotary machinery based on information entropy and a probabilistic neural network. *Sensors* **2018**, *18*, 337. [CrossRef] [PubMed]
22. Zhang, Z.; Han, H.; Cui, X.; Fan, Y. Novel application of multi-model ensemble learning for fault diagnosis in refrigeration systems. *Appl. Therm. Eng.* **2020**, *164*, 114516. [CrossRef]
23. Quan, H.; Li, J.; Peng, D. Multisensor fault diagnosis based on data fusion using D-S theory. In Proceedings of the 33rd Chinese Control Conference (CCC), Nanjing, China, 28 July 2014; IEEE: New York, NY, USA, 2014.
24. Li, X.; Jiang, H.; Niu, M.; Wang, R. An enhanced selective ensemble deep learning method for rolling bearing fault diagnosis with beetle antennae search algorithm. *Mech. Syst. Signal Processing* **2020**, *142*, 106752. [CrossRef]
25. Basir, O.; Yuan, X. Engine fault diagnosis based on multi-sensor information fusion using Dempster–Shafer evidence theory. *Inf. Fusion* **2007**, *8*, 379–386. [CrossRef]
26. Zhao, K.; Sun, R.; Li, L.; Hou, M.; Yuan, G.; Sun, R. An improved evidence fusion algorithm in multi-sensor systems. *Appl. Intell.* **2021**, *51*, 7614–7624. [CrossRef]
27. Dragomiretskiy, K.; Zosso, D. Variational Mode Decomposition. *IEEE Trans. Signal Processing* **2014**, *62*, 531–544. [CrossRef]
28. Dibaj, A.; Ettefagh, M.M.; Hassannejad, R.; Ehghaghi, M.B. A hybrid fine-tuned VMD and CNN scheme for untrained compound fault diagnosis of rotating machinery with unequal-severity faults. *Expert Syst. Appl.* **2021**, *167*, 114094. [CrossRef]
29. Tong, J.; Luo, J.; Pan, H.; Zheng, J.; Zhang, Q. A Novel Cuckoo Search Optimized Deep Auto-Encoder Network-Based Fault Diagnosis Method for Rolling Bearing. *Shock. Vib.* **2020**, *2020*, 1–12. [CrossRef]

30. Shao, H.; Xia, M.; Wan, J.; de Silva, C.W. Modified stacked auto-encoder using adaptive morlet wavelet for intelligent fault diagnosis of rotating machinery. *IEEE/ASME Trans. Mechatron.* **2021**, *27*, 24–33. [CrossRef]
31. Shao, H.; Jiang, H.; Zhao, H.; Wang, F. A novel deep autoencoder feature learning method for rotating machinery fault diagnosis. *Mech. Syst. Signal Processing* **2017**, *95*, 187–204. [CrossRef]
32. Tanjung, J.P. Classification of wheat seeds using neural network backpropagation algorithm. *J. Inform. Telecommun. Eng.* **2021**, *4*, 335–342. [CrossRef]
33. Van der Maaten, L.; Hinton, G. Visualizing data using t-SNE. *J. Mach. Learn. Res.* **2008**, *9*, 2579–2605.

Article

Research on Fault Feature Extraction Method Based on Parameter Optimized Variational Mode Decomposition and Robust Independent Component Analysis

Jingzong Yang [1],*, Chengjiang Zhou [2],* and Xuefeng Li [3,4]

[1]   School of Dig Data, Baoshan University, Baoshan 678000, China
[2]   School of Information, Yunnan Normal University, Kunming 650500, China
[3]   College of Automobile and Traffic Engineering, Nanjing Forestry University, Nanjing 210037, China; lixuefengseu@foxmail.com
[4]   School of Transportation, Southeast University, Nanjing 211189, China
*   Correspondence: 171847260@masu.edu.cn (J.Y.); chengjiangzhou@foxmail.com (C.Z.)

Citation: Yang, J.; Zhou, C.; Li, X. Research on Fault Feature Extraction Method Based on Parameter Optimized Variational Mode Decomposition and Robust Independent Component Analysis. *Coatings* **2022**, *12*, 419. https://doi.org/10.3390/coatings12030419

Academic Editor: Alessandro Latini

Received: 30 January 2022
Accepted: 15 March 2022
Published: 21 March 2022

Publisher's Note: MDPI stays neutral with regard to jurisdictional claims in published maps and institutional affiliations.

Abstract: The variational mode decomposition mode (VMD) has a reliable mathematical derivation and can decompose signals adaptively. At present, it has been widely used in mechanical fault diagnosis, financial analysis and prediction, geological signal analysis, and other fields. However, VMD has the problems of insufficient decomposition and modal aliasing due to the unclear selection method of modal component k and penalty factor α. Therefore, it is difficult to ensure the accuracy of fault feature extraction and fault diagnosis. To effectively extract fault feature information from bearing vibration signals, a fault feature extraction method based on VMD optimized with information entropy, and robust independent component analysis (RobustICA) was proposed. Firstly, the modal component k and penalty factor α in VMD were optimized by the principle of minimum information entropy to improve the effect of signal decomposition. Secondly, the optimal parameters weresubstituted into VMD, and several intrinsic mode functions (IMFs) wereobtained by signal decomposition. Secondly, the kurtosis and cross-correlation coefficient criteria were comprehensively used to evaluate the advantages and disadvantages of each IMF.And then, the optimal IMFs were selected to construct the observation signal channel to realize the signal-to-noise separation based on RobustICA. Finally, the envelope demodulation analysis of the denoised signal was carried out to extract the fault characteristic frequency. Through the analysis of bearing simulation signal and actual data, it shows that this method can extract the weak characteristics of rolling bearing fault signal and realize the accurate identification of fault. Meanwhile, in the bearing simulation signal experiment, the results of kurtosis value, cross-correlation coefficient, root mean square error, and mean absolute error are 6.162, 0.681, 0.740, and 0.583, respectively. Compared with other traditional methods, better index evaluation value is obtained.

Keywords: variational mode decomposition (VMD); information entropy; robust independent component analysis (RobustICA); fault feature extraction; rolling bearing

1. Introduction

Rolling bearing is one of the widely used parts in rotating machinery and equipment in the manufacturing system. It widely exists in the power end, transmission end, and execution end. Due to its complex working environment, it is very prone to failure. According to relevant statistics, about 30% of failures in rotating machinery are related to rolling bearings. Therefore, the research on fault diagnosis of rolling bearing is of great significance to ensure the production safety of relevant enterprises. However, in the actual operation process, due to the influence of external noise, receiving distance, and sensor working environment, the fault characteristics of this component are submerged in the interference of intensebackground noise, which has a significantimpact on fault diagnosis.

At present, the bearing fault diagnosis method based on signal processing has experienced three processes: time–domain analysis, frequency domain analysis, and time–frequency domain analysis. Because the fault signal of rolling bearing is non-stationary and nonlinear, the time–frequency domain analysis method is relatively suitable for processing this signal. Since the development of the time–frequency analysis method, the algorithms often used by relevant experts and scholars include Short-time Fourier transform (STFT) [1,2], S-transform [3], Wigner Ville distribution (WVD) [4], wavelet transform (WT) [5,6], etc., singular spectrum decomposition (SSD) [7], spectral kurtosis (SK) [8], morphological filtering (MF) [9], and singular spectrum analysis (SSA) [10], etc. However, these methods havetheir ownlimitations. For example, the window function of STFT is fixed, which is not conducive to the analysis of non-stationary bearing fault signals. The standard deviation of the S-transform in the Gaussian window function is fixed as the reciprocal of the frequency, which makes the time–frequency aggregation of the high-frequency part of the signal not ideal.WVD has good time–frequency aggregation, but there is cross-term interference. Although WT has the time–frequency local analysis ability of adjustable window, the scale of wavelet transform does not have a good corresponding relationship with the frequency of signal. In SK algorithm, how to set the parameters of passband center frequency and resonance bandwidth will affect its application effect. In MF algorithm, it is difficult to effectively select the type and size of the structuring element. In SSA algorithm, the embedding dimension and time delay of phase space reconstruction cannot be set automatically. After continuous development, relevant scholars have proposed many adaptive signal processing methods based on previous research results and applied them to the fault feature extraction of rolling bearing. For example, empirical mode decomposition (EMD) [11], ensemble empirical mode decomposition (EEMD) [12], local mean decomposition (LMD) [13], etc., Jiang et al. [14] proposed a fault feature extraction method combining EMD and 1.5-dimensional spectrum. In this method, the signal components decomposed by EMD are screened and reconstructed Then the reconstructed Hilbert envelope signal is analyzed by a 1.5-dimensional spectrum to obtain the characteristic fault frequency of bearing. The effectiveness of the proposed method is proved by experimental analysis. Fan et al. [15] used EMD and pesudo-Wigner-Ville distribution (PWVD) to convert rolling bearing vibration signals with different fault degrees into contour time–frequency images, then extracted energy distribution values as features, and constructed a fault diagnosis model based onfuzzy c-means (FCM). Experimental results show that this method has high recognition accuracy. Hou et al. [16] proposed a fault diagnosis method composed of EEMD, permutation entropy (PE), and Gath Geva (GG) clustering algorithm to solve the problem that it is difficult to identify the fault type of rolling bearing. Experimental results show that the proposed fault diagnosis method can achieve better clustering results. Liang et al. [17] proposed a fault diagnosis method based on long-term and short-term memory network (LSTM) and LMD, to improve the defects of the LMD method. The results show that the method successfully extracts the characteristic frequencies of rolling bearing. Although the above time–frequency analysis methods have achieved certain results, they are based on the principle of recursive decomposition, so they have not been well solved in the aspects of modal aliasing and endpoint effect. To solve this problem, Dragomiretskiy et al. [18] proposed a new adaptive decomposition method variational modal decomposition. The algorithm makes the decomposition result stable through the construction of the variational problem and is applied to the fault diagnosis of rotating machinery. Ye et al. [19] decomposed the bearing vibration signal by the VMD method and introduced the characteristic capability ratio criterion to screen the qualified signal components for reconstruction. Then, the multi-dimensional features of the signal are extracted and input into the particle swarm optimization (PSO) andsupport vector machine (SVM) classification model for fault diagnosis. The results show that the proposed method has higher recognition accuracy than the existing methods. Li et al. [20] proposed a fault diagnosis method combining VMD and fractional Fourier transform (FRFT) to solve the problem that it is difficult to extract fault features and over decomposition when applying the VMD method to rolling bearing fault diagnosis. By analyzing the results of simulation experiments, the method has a good effect. Xing et al. [21] combined VMD, Tsallis entropy, and fuzzy c-means clustering (FCM) and applied them to fault diagnosis. Through the analysis of the measured vibration signal of rolling bearing, the results show that this method can obtain better results than EMD and LMD methods.

The signal processed by time–frequency analysis contains a lot of noise, which has a specificimpact on the fault feature extraction. Therefore, the signal needs further processing. In recent years, the technology based on blind source separation has become one of the research hotspots. This technology optimizes multiple observation signals according to the principle of statistical independence and decomposes them into several independent components, so as to achieve the purpose of signal enhancement. Robust independent component analysis (RobustICA), as an algorithm with outstanding advantages in blind source separation methods, has been widely used in signal analysis, fault diagnosis, and other fields because of its good effect in signal-to-noise separation effect and calculation efficiency [22–24]. Yang et al. [25] proposed a signal noise-reduction method based on the combination of complementary ensemble empirical mode decomposition (CEEMD) and RobustICA to reduce the noise of pipeline blockage signals. Through the processing and analysis of simulation signals and pipeline blockage detection signals, the results verify the effectiveness of the proposed method. Yao et al. [26] studied the noise reduction of internal combustion engine signals by using the combination of Gammatone filter bank and RobustICA. Experiments show that the classification effect of signal and noise obtained by this method is good. Zhao et al. [27] combined EEMD, RobustICA, and Prony algorithms and applied them to the identification of low-frequency oscillation signals in the power system. Experiments show that the proposed method has a strong anti-interference ability and can effectively suppress noise.

In this paper, a fault feature extraction method based on VMD optimized with information entropy and RobustICA is proposed. Firstly, the fault signal is decomposed by VMD, and the number of modal components k and penalty parameters α are selected according to the optimization principle of minimum information entropy. Then, the optimal parameters are substituted into VMD and the signal decomposition operation is carried out. Secondly, the signal components are filtered through the constructed signal component screening criteria, and the observation signal channel is constructed, so as to realize the signal-to-noise separation based on RobustICA. Finally, the denoised signal is demodulated by Hilbert envelope, and the fault characteristic frequency is extracted.

The main contributions of this paper are summarized as follows:

(1) A method of optimizing VMD by information entropy is proposed to set initialization parameters so that VMD can adaptively expand signal decomposition;
(2) In this study, kurtosis and cross-correlation coefficient criteria are comprehensively used to evaluate the advantages and disadvantages of each eigenmode function, and the optimal eigenmode function is selected to construct the observation signal channel, which completes the separation of useful signal and noise signal; and
(3) The effectiveness and feasibility of the proposed method are verified by experimental analysis using simulation signals and actual bearing data sets.

The structure of this paper is as follows. Section 2 introduces the basic principles of VMD, information entropy, and the RobustICA algorithm. Section 3 introduces the specific implementation process of the fault feature extraction method based on information entropy optimization VMD and RobustICA. In Section 4, the stimulation signal is experimentally studied by using the proposed method. In Section 5, the effect of the method is further verified by the actual bearing fault signal experiment. Section 6 is the discussion and conclusions.

2. Basic Methods

2.1. VMD

Variational modal decomposition (VMD) is developed by University of California scholars Dragomiretskiy et al., in 2014 [18]. Based on Wiener filtering, this method searches for the optimal solution of the input signal within the framework of variational model. It can adaptively update the center frequency, bandwidth, and corresponding sub signals, and decompose the independent components of the signal from the frequency domain. As a non-recursive signal analysis method, the core idea of the VMD method is to determine the intrinsic mode function (IMF) by solving the variational problem. Therefore, in the VMD algorithm, the IMF component obtained by signal decomposition is different from that in EMD and LMD algorithms. The original signal is non recursively decomposed into several IMF components with limited bandwidth:

$$\mu_k(t) = A_k(t)\cos[\phi_k(t)] \tag{1}$$

In the above formula, $A_k(t)$ is the instantaneous amplitude of $\mu_k(t)$, and $A_k(t) \geq 0$. $\phi_k(t)$ is phase and $\phi_k(t) \geq 0$. $\omega_k(t)$ is the instantaneous phase of $\mu_k(t)$.

$$\omega_k(t) = \phi_{k}{}'(t) = \frac{d\phi_k(t)}{dt} \tag{2}$$

In the above formula, $\mu_k(t)$ can be regarded as a harmonic signal, its amplitude is $A_k(t)$ and its frequency is $\omega_k(t)$.

It is assumed that each mode has a limited bandwidth with central frequency, and the central frequency and bandwidth will be updated continuously in the decomposition process. Then the variational problem can be expressed as finding r modal functions $\mu_k(t)$ and minimizing the estimated bandwidth forthe sum of all modal functions. The sum of modes is the input signal.VMD algorithm can obtain k discrete modes $\mu_k(t)$ ($r \in 1, 2, \cdots, R$) by decomposing signal $X(t)$. Then, the frequency bandwidth of each modal signal is estimated in the following manner.

(1) Hilbert transform is extended to the modal function, and the marginal spectrum is obtained.

$$\left(\delta(t) + \frac{j}{\pi t}\right) * \mu_k(t) \tag{3}$$

(2) Each estimated center band is modulated to the corresponding fundamental band.

$$\left[(\delta(t) + \frac{j}{\pi t}) * \mu_k(t)\right]e^{-j\omega_k t} \tag{4}$$

(3) The square L^2 norm of the demodulated signal gradient is obtained.

$$\left\| d_t[(\delta(t) + \frac{j}{\pi t}) * \mu_k(t)]e^{-j\omega_k t} \right\|_2^2 \tag{5}$$

By constructing the VMD variational constraint model based on the above formula, the following formula can be formed.

$$\begin{cases} \min_{\{\mu_k\},\{\omega_k\}} \left\{ \sum_{k=1}^{k} \|d_t[(\delta(t) + \frac{j}{\pi t}) * \mu_k(t)]e^{-j\omega_k t}\|_2^2 \right\} \\ \text{s.t. } \sum_{k=1}^{k} \mu_k(t) = x(t) \end{cases} \tag{6}$$

In the above formula, $\{\mu_k(t)\} = \{\mu_1, \mu_2, \mu_3, \cdots \mu_k\}$ is the function set of each mode. $\{\omega_k\} = \{\omega_1, \omega_2, \omega_3, \cdots \omega_k\}$ is the central frequency set. $\delta(t)$ is the unit pulse function. d_t is the derivative of the function over time t. s.t. is the constraint. $x(t)$ is the original input signal, where k is the number of decompositions.

By introducing penalty factor and Lagrange multiplication operator b, the constrained model problem in Equation (5) can be transformed into a non-constrained model problem, as in Equation (6).

$$L(\{\mu_k\},\{\omega_k\},\lambda) = \alpha \sum_{k=1}^{k} \left\| d_t\left[(\delta(t) + \frac{j}{\pi t}) * \mu_k(t)\right]e^{-j\omega_k t} \right\|_2^2 + \left\| x(t) - \sum_{k=1}^{k} \mu_k(t) \right\|_2^2 + \left\langle \lambda(t), x(t) - \sum_{k=1}^{k} \mu_k(t) \right\rangle \quad (7)$$

By constantly iteratively searching for the minimum point of Lagrange function L, the original input signal a will be decomposed into k modal functions $\mu_k(t)$.

When using VMD decomposition algorithm to adaptively decompose the signal, the decomposition parameters need to be set in advance. Theoretical research shows that the parameters that have a great impact on the decomposition effect mainly include the number of decomposition k and the penalty parameter α. Therefore, setting these two parameters only by experience will bring great errors to the decomposition results of VMD. Among them, the size of k value is directly related to the decomposition effect of VMD. If the value of k is too small, it will lead to under decomposition of the signal, and the resulting signal is not completely decomposed into components. If the value of k is too large, the signal will be decomposed too much, resulting in over decomposition. At the same time, the value of a has a certain impact on the bandwidth of the decomposed component. If the value of a is too small, the bandwidth of the decomposed component will be too large, and some components will include other components. If the value of a is too large, the bandwidth of the decomposed component will be too small, and some components in the decomposed signal may be lost.

2.2. Information Entropy

The concept of entropy originates from thermodynamics and is used to describe the disorder degree of the system. Based on this idea, scholar Shannon proposed the concept of information entropy [28]. Information entropy is a concept used to measure the amount of information in information theory. When a system is more orderly, the value of information entropy will be smaller. On the contrary, when the disorder degree of the system becomes higher, the value of information entropy will become larger. The calculation formula of information entropy is as follows:

$$G(x) = -\sum_{i=1}^{n} P(x_i)\log P(x_i) \quad (8)$$

where $P(x_i)$ represents the probability of occurrence of an event x_i in the system, and n represents the number of samples to be analyzed.

2.3. Robust Independent Component Analysis

RobustICA algorithm is proposed by Zarzoso et al. [29,30]. Because the observation data used in it does not need whitening preprocessing, it can only meet the condition that its mean value is zero, so the problem of introducing error is avoided. The algorithm realizes the kurtosis optimization by using linear search and algebraic calculation of the global optimal step size. The frame of independent component analysis is shown in Figure 1 [31]. Assuming that the mixed data containing noise is X and the output signal is $Y = WX$, the kurtosis formula can be expressed as follows:

$$K(W) = \frac{E\{|Y|^4\} - 2E^2\{|Y|^2\} - |E\{Y^2\}|^2}{E^2\{|Y|^2\}} \quad (9)$$

where $E\{\cdot\}$ represents mathematical expectation, W represents separation matrix.

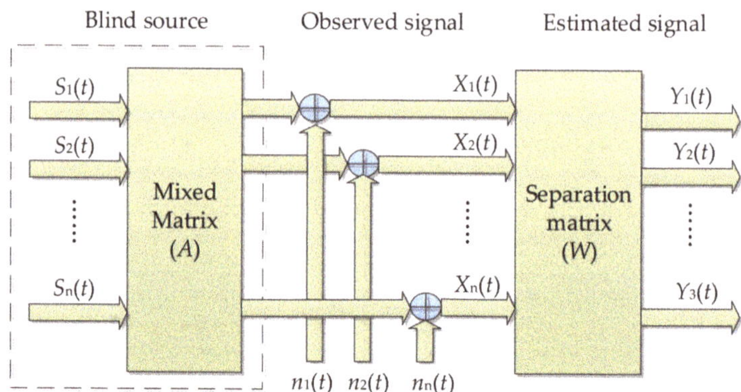

Figure 1. Frame of independent component analysis.

An exact linear search is performed by using the absolute value of kurtosis as the objective function:

$$\mu_{opt} = \underset{u}{\mathrm{argmax}}|K(W + \mu g)| \tag{10}$$

The search direction g is usually gradient, namely. $g = \nabla_w K(W)$. It is expressed as follows:

$$g = \nabla_w K(W) = \frac{4}{E^2\{|Y|^2\}}\left\{E\{|Y^2|\}YX - E\{YX\}E\{Y^2\} - \frac{(E\{|Y|^4\} - |E\{Y^2\}|^2)E\{YX\}}{E\{|Y|^2\}}\right\} \tag{11}$$

where K indicates kurtosis. $E\{\cdot\}$ represents mathematical expectation. ∇_w indicates gradient.

In the process of each iterative operation, the operation steps of RobustICA is as follows:

(1) Find the coefficients of the optimal step polynomial:

$$P(u) = \sum_{k=0}^{4} a_k u^k \tag{12}$$

(2) Extract the root of the optimal polynomial (12).

(3) In the search direction, the root value that maximizes the kurtosis is selected as the optimal step

$$\mu_{opt} = \underset{u}{\mathrm{argmax}}|K(W + \mu g)| \tag{13}$$

(4) Update the separation vector: $W^+ : W^+ = W + \mu_{opt}g;$

(5) Normalized the separation vector: $W^+ : W^+ \leftarrow \frac{W^+}{\|W^+\|};$

(6) If there is no convergence, return to Step (1), otherwise the solution of the separation vector is completed.

3. Fault Feature Extraction Based on VMD Optimized with Information Entropy and RobustICA

3.1. Parameter Optimization of VMD

During the working process of rolling bearing, because the inner ring will rotate with the shaft, the pressure of therolling bearing changes periodically. When the rolling bearing fails, the damaged part's surface will contact other parts and collide. Therefore, it will produce periodic pulse impact. According to this bearing characteristic, the information entropy theory can be used to measure the above changes. When the periodic impact is more uniform, the signal is more orderly, so the value of information entropy will be smaller. Therefore, if the IMF component obtained by VMD decomposition contains more fault information, its performance will be more orderly, and the value of information entropy will be smaller. In this research, envelope entropy is introduced to measure signal sparsity. The envelope entropy of $IMF_i(j)$ component of signal decomposed by VMD algorithm can be expressed as follows:

$$E_i = -\sum_{j=1}^{n} P_{i,j} \log P_{i,j} \tag{14}$$

$$P_{i,j} = a_i(j) / \sum_{j=1}^{n} a_i(j) \tag{15}$$

where, i represents the sequence number of IMF obtained by the decomposition of signal $x(j)$ (i = 1, 2, 3, $\cdots$). $P_{i,j}$ represents the normalized result of $a_i(j)$. $a_i(j)$ is the envelope signal of signal component $IMF_i(j)$ after Hilbert envelope demodulation. Based on this, this paper adopts the principle of minimum envelope entropy to determine the initialization parameters of VMD.

In parameter optimization, the value of α can be given first, and the optimal value of mode k can be determined based on the principle of minimum envelope entropy. After obtaining the value of k, the optimal value of α is further determined based on the value of k. Its optimization objective can be expressed as shown in Formula (16). Where k_{min} and k_{max} represent the minimum and maximum values of the interval range of modal k search, respectively. α_{min} and α_{max} represent the minimum and maximum values of the interval range searched by the penalty factor α, respectively. The specific optimization process is shown in Figure 2.

$$\begin{cases} \min\{E\} \\ s.t. \ k_{min} \leq k_z \leq k_{max} \\ s.t. \ \alpha_{min} \leq \alpha_z \leq \alpha_{max} \end{cases} \tag{16}$$

The specific steps are as follows:

(1) Optimize the modal number k. First, set the initial value of mode number k to 3, and use the default value of 2500 for the penalty factor α. Then, the fault signal is decomposed by VMD, and the envelope entropy of all modes is obtained according to the calculation formula of envelope entropy. According to the calculation results, judge whether the envelope entropy obtainedthe minimum value. If it is the minimum value, select the value at this time as the optimal value. If not, perform $k + 1$ operation on the value of a modal number, and continue to repeat the above analysis steps until the minimum envelope entropy value is found. Then record the corresponding k value. In this research, the search range of k is from to 3 to 15, and the step size is 1.

(2) Optimize the penalty factor. Based on the value of k determined in the previous step to further determine the optimal value. The search principle is the same as the previous step. When the minimum envelope entropy value is found, the search is terminated. Otherwise, perform the operation of $\alpha + 50$ and repeat the above analysis steps until the smallest value of envelope entropy is found, and then record the corresponding value. In this research, the search range of α is set from 100 to 2500, and the step size is 50.

(3) Based on the optimization results of the first and second steps, the optimal values of k and α are substituted into VMD, and the parameters are initialized. Then perform VMD to get the optimal IMF component.

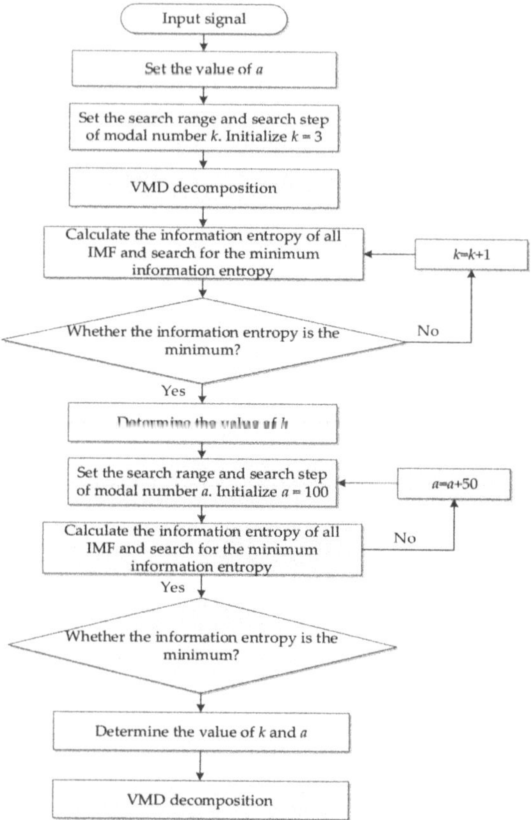

Figure 2. Parameter optimization process of VMD.

3.2. Screening Criteria for IMF Components

As a dimensionless parameter, kurtosis is often used for the distribution characteristics of vibration signals. For the IMF signal components obtained by VMD decomposition, when the signal component's kurtosis value is more prominent, it contains more impact components. At the same time, the cross-correlation coefficient represents the correlation between signals. The greater the correlationcoefficient for the IMF signals components, the more sensitive information it contains. Conversely, the more interference components it contains. Based on this, in the fault signal processing of rolling bearings, this research will comprehensively use the above two criteria to determine which signal components are used to construct the observation signal channel of the RobustICA, and then achieve the purpose of noise reduction. In selecting signal components, the principle adopted is that the correlation number is more significant than 0.3 and the kurtosis value is greater than 3. In this way, it is possible to avoid the problem of using a single index to select only the signal component with the most significantkurtosis value, which leads to the loss of sensitive information of part of the fault signal. The formula for calculating kurtosis and cross-correlation coefficient is as follows:

$$\text{Kurtosis}_i = \frac{1}{\mu} \sum_{i=1}^{n} \left(\frac{\theta_i - \overline{x}}{\updownarrow} \right)^4 \tag{17}$$

In the original vibration signal, θ_i and $\overline{x}$ are its actual value and average value. $\updownarrow$ is the standard deviation. μ is the number of samples.

$$Correlation = \frac{\sum\limits_{i=0}^{n} (\theta_i - \overline{x})(\delta_i - \overline{y})}{\sqrt{\sum\limits_{i=0}^{n} (\theta_i - \overline{x})^2(\delta_i - \overline{y})^2}} \tag{18}$$

In the original vibration signal, θ_i and $\overline{x}$ its specific value and average value. Meanwhile, δ_i and $\overline{y}$ are the specific and average values of signal ϑ.

3.3. Algorithm Steps and Flow

The specific steps of the fault feature extraction method based on information entropy optimization VMD and RobustICA are as follows. The fault feature extraction process is shown in Figure 3.

(1) The vibration signal of rolling bearing is collected, and the VMD is optimized by information entropy to find the optimal value of initialization parameter k and α.
(2) The obtained optimal parameters k and α are substituted into VMD, and the relevant parameters are initialized. Secondly, the signal is decomposed into several IMF components by VMD decomposition.
(3) Kurtosis criterion and cross-correlation coefficient criterion are comprehensively used to evaluate the advantages and disadvantages of each IMF component, and the optimal IMF component is selected to construct the observation signal channel.
(4) The observed signal and virtual noise channel signal are separated by RobustICA algorithm, and the useful signal is obtained.
(5) Envelope demodulation is performed on the signal after RobustICA noise reduction, and the fault characteristic frequency is extracted. Then it is compared with the theoretical value of bearing fault characteristics to identify the fault category.

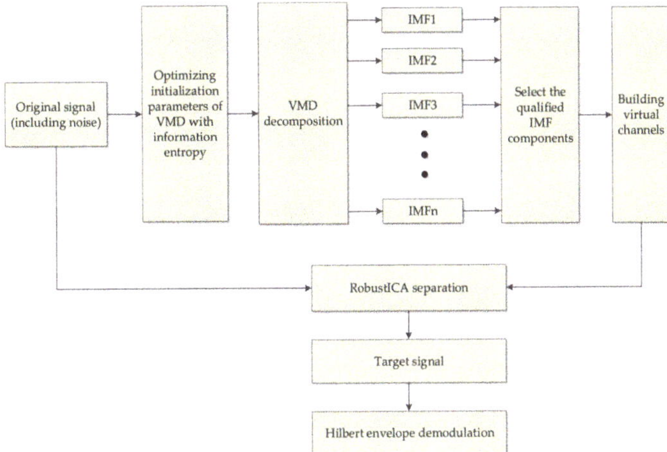

Figure 3. Fault feature extraction method based on VMD optimized with information entropy and RobustICA.

4. Simulations and Comparative Analysis

Bearing is one of the most commonly used general parts in all kinds of rotating machinery. It plays a role in bearing and transmitting load in mechanical equipment, and it is very prone to failure. Therefore, to verify the performance of the algorithm proposed in this paper, a typical model was used to simulate the periodic impact signal caused by bearing fault [32]. Firstly, a set of periodic pulse signals was simulated to simulate the fault impact signal, and on this basis, Gaussian white noise was added to the fault impact signal to simulate the bearing fault vibration signal polluted by environmental noise. In this study, Matlab (version R2009a) was selected as the vibration signal modeling and simulation software to build the signal model. The expression of the simulated signal is as follows:

$$s(t) = y_0 e^{-2\pi f_n \xi t} \sin(2\pi f_n \sqrt{1 - \xi^2} t) \tag{19}$$

In the above formula: carrier frequency f_n = 3000 Hz, displacement constant y_0 = 5, damping coefficient ξ = 0.1, period T = 0.01 s, sampling frequency f_s = 20 KHz, and number of sampling points n = 4096, where t represents the sampling time. Through calculation, it can be seen that the fault frequency f_0 = 100 Hz. In order to simulate the noise interference of rolling bearing during operation, SNR = −5 dB white Gaussian noise was added to the original signal $s(t)$.

The time–domain waveform of the original signal is shown in Figure 4. After adding noise, the time–domain waveform and frequency–domain waveform of the mixedsignal after adding noise is shown in Figures 5 and 6. Analyzing the above diagrams showsthat most of the impact signal features were covered up under the interference of background noise, which brings some difficulty to the fault feature extraction.

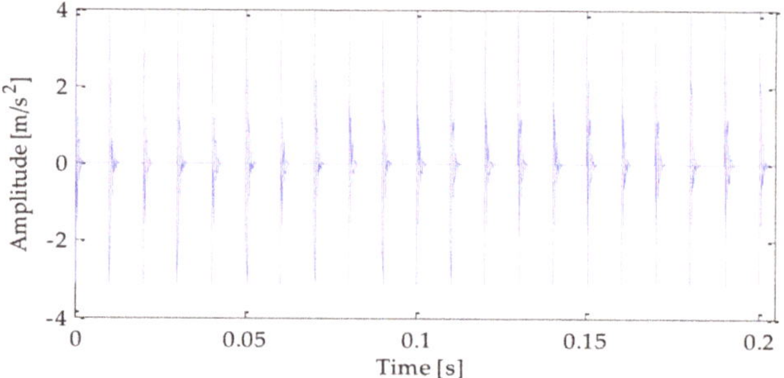

Figure 4. Time–domain waveform of the simulated signal.

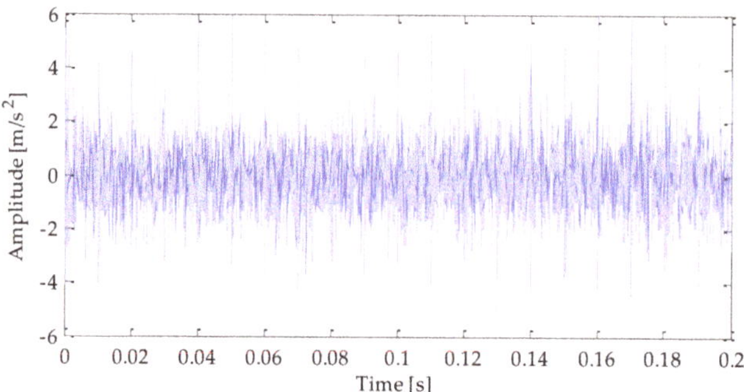

Figure 5. Time–domain waveform of the mixed signal.

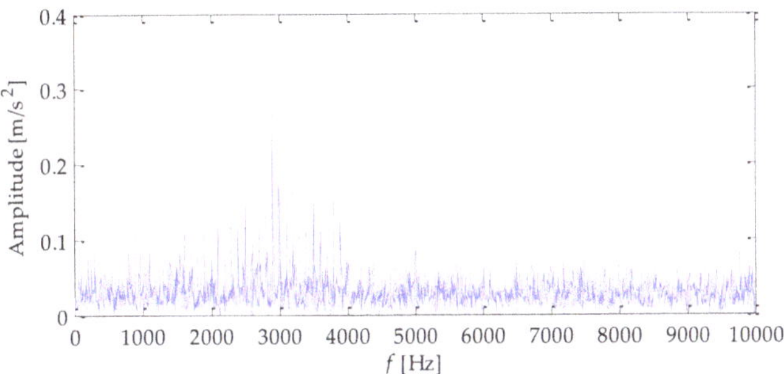

Figure 6. Frequency–domain waveform of the mixed signal.

Since the mixed signal was severely interfered with by noise, next, the signal will be decomposed by VMD. Before VMD decomposition, information entropy must be used to optimize VMD to determine the parameters k and α in VMD. Firstly, our experiment used the default value a, which was 2500. Meanwhile, we initialized $k = 3$, and the search range of k was set to [3,15]. The value of optimal mode k was searched according to the principle of minimum envelope spectral entropy. The relationship between k and envelope spectral entropy is shown in Figure 7. From the transformation trend of the value of k in the figure, it can be seen that with the increasing value of k, the corresponding envelope spectral entropy was also increasing. Therefore, $k = 3$ was taken as the optimal value.

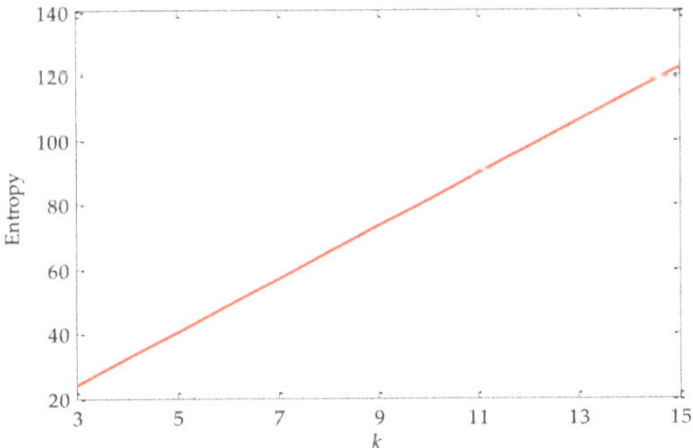

Figure 7. Curve of fitness varying with k value.

After selecting the value of the optimal mode K, the value of α was then initialized. The search range was set to [100, 2500]. Similarly, the value of the optimal mode α was searched according to the principle of minimum envelope spectral entropy. The relationship with envelope spectral entropy is shown in Figure 8. From the transformation trend of the value of α in the figure, it can be seen that with the continuous increase of α, the value of the corresponding envelope spectral entropy was decreasing and gradually tends to be flat. When the value of α was 2500, the optimal value can be obtained. Therefore, after parameter optimization, the selected optimal parameter combination K and α was [3, 2500].

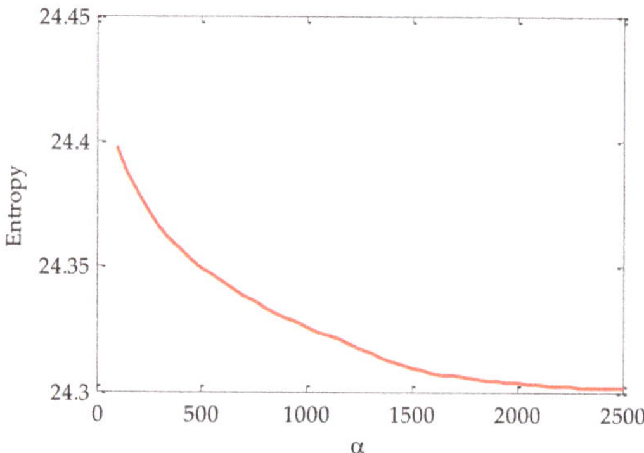

Figure 8. Curve of fitness varying with α value.

Next, we performed VMD decomposition on the mixed signal. After VMD decomposition optimized by information entropy, three IMF components were obtained. The decomposition results are shown in Figure 9. In order to compare the effects of different methods, the traditional LMD decomposition method, EMD decomposition method, and EEMD decomposition method were used for the time–frequency analysis of mixed signals. The signal decomposition results obtained based on the above three methods are shown in Figure 10a,b. Figures 10 and 11 showed that there were more signal components decomposed by LMD, EMD and EEMD methods, and the signal components obtained by EMD and EEMD had certain modal aliasing and endpoint effect. At the same time, faultycomponents were generated.

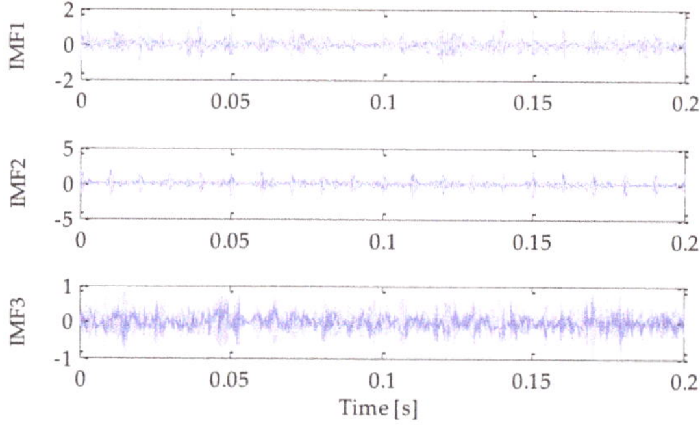

Figure 9. VMD decomposition result.

To select the appropriate signal component from the decomposition results obtained by the above methods, this experiment will combine kurtosis and cross-correlation coefficient to select. Firstly, all signal components' correlation coefficient $C(t)$ and kurtosis value $Q(t)$ were calculated. The calculated results are shown in Tables 1–4. It can be seen from Table 1 that the IMF1 component and IMF2 component obtained by VMD decomposition meet the conditions that the correlation value was more significant than 0.3 and the kurtosis value was greater than 3. This shows that the correlation between the above two signal components and the original signal was high, and the signal contained more impact components. Therefore, the IMF1 and IMF2 components were selected to reconstruct the observation signal channel. Secondly, it can be seen from Table 2 that the PF1 component and PF2 component obtained by LMD decomposition met the conditions that the correlation value is more significantthan 0.3 and the kurtosis value is greater than 3. Therefore, PF1 component and PF2 component were selected to reconstruct the observation signal channel.

Meanwhile, it can be seen from Table 3 that the IMF1 component and IMF2 component obtained by EMD decomposition met the conditions that the correlation value is more significant than 0.3 and the kurtosis value is greater than 3. Therefore, IMF1 and IMF2 components were selected to reconstruct the observation signal channel. It can be seen from Table 4 that the IMF1 component, IMF2 component and IMF3 component obtained by EEMD decomposition meet the conditions that the correlation value is more significant than 0.3 and the kurtosis value is greater than 3. Therefore, the above three signal components were selected to reconstruct the observation signal channel, and the remaining signal components were used to reconstruct the noise signal channel. Finally, on this basis, the RobustICA algorithm was used to separate signal and noise. The noise-reduction results obtained by using the method proposed in this paper and LMD–RobustICA, EMD–RobustICA, and EEMD–RobustICA are shown in Figures 12–15.

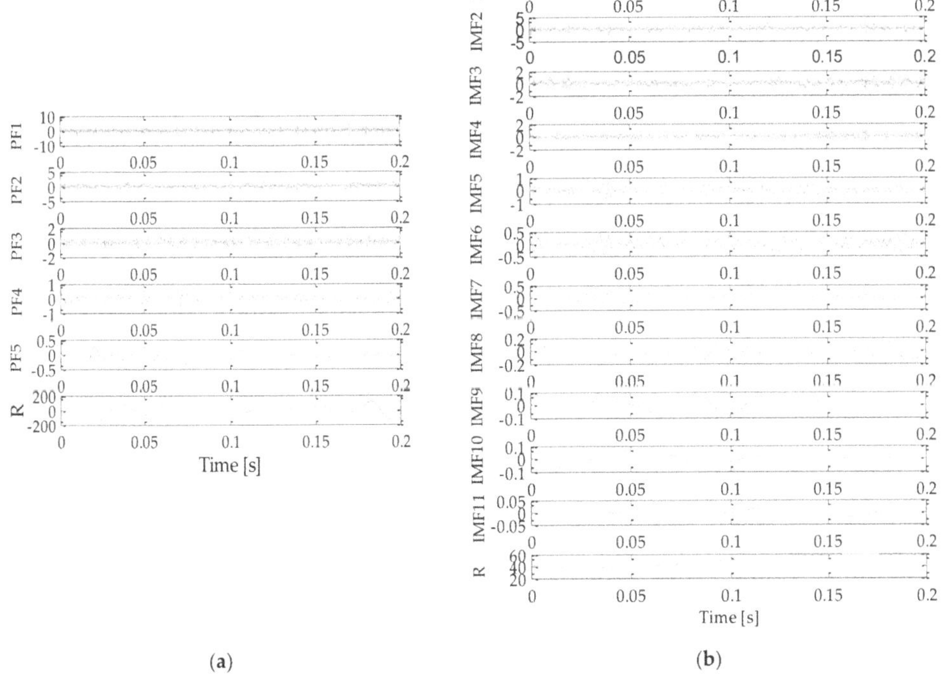

(a)

(b)

Figure 10. LMD (**a**) and EMD (**b**) decomposition result.

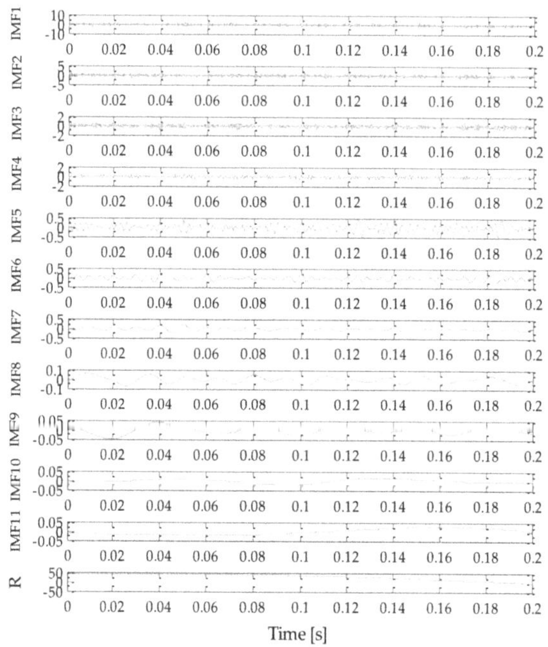

Figure 11. EEMD decomposition result.

Table 1. Index values of different IMF signal components (VMD).

Parameter	IMF1	IMF2	IMF3
$C(t)$	0.433	0.575	0.389
$Q(t)$	4.402	6.221	2.916

Table 2. Index values of different PF signal components (LMD).

Parameter	PF1	PF2	PF3	PF4	PF5
$C(t)$	0.835	0.418	0.222	0.111	0.062
$Q(t)$	4.630	4.161	3.572	4.207	3.535

Table 3. Index values of different IMF signal components (EMD).

Parameter	IMF1	IMF2	IMF3	IMF4	IMF5	IMF6
$C(t)$	0.727	0.516	0.283	0.191	0.139	0.107
$Q(t)$	3.606	4.216	2.833	3.124	2.859	3.116
Parameter	IMF7	IMF8	IMF9	IMF10	IMF11	R
$C(t)$	0.057	0.038	0.018	0.001	0.009	0.009
$Q(t)$	4.671	2.318	2.316	3.145	1.505	1.323

Table 4. Index values of different IMF signal components (EEMD).

Parameter	IMF1	IMF2	IMF3	IMF4	IMF5	IMF6
$C(t)$	0.791	0.587	0.353	0.238	0.189	0.127
$Q(t)$	4.069	3.981	3.248	3.662	3.072	2.818
Parameter	**IMF7**	**IMF8**	**IMF9**	**IMF10**	**IMF11**	**R**
$C(t)$	0.077	0.045	0.020	0.009	0.013	-0.001
$Q(t)$	3.503	2.439	2.079	2.239	1.463	2.655

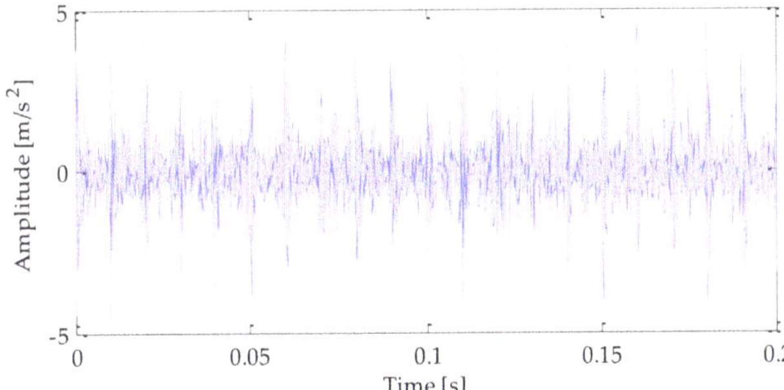

Figure 12. Noise-reduction results by the method proposed in this paper.

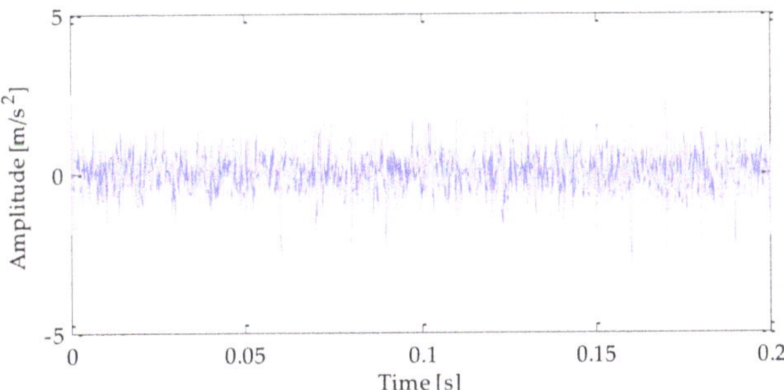

Figure 13. LMD–RobustICA noise-reduction results.

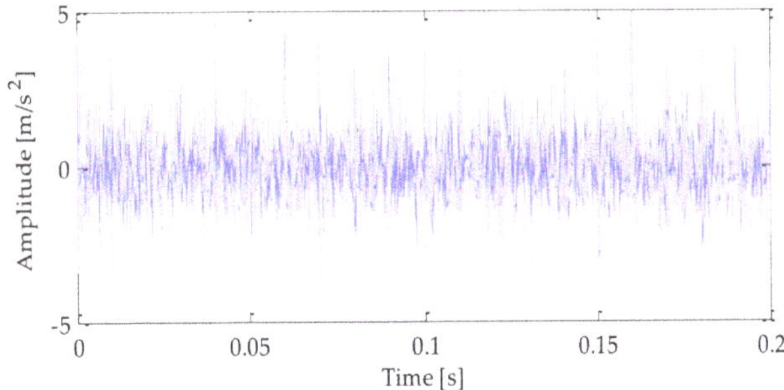

Figure 14. EMD–RobustICAnoise-reduction results.

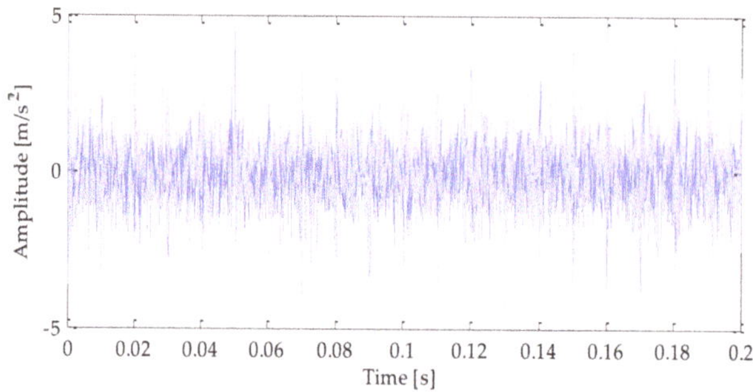

Figure 15. EEMD–RobustICAnoise-reduction results.

By analyzing the noise-reduction results in Figures 12–15, it can be seen that after the noise-reduction method proposed in this article, the impact components in the signal have been revealed. In contrast, the results obtained by the LMD–RobustICA, EMD–RobustICA, and EEMD–RobustICA were not significant. To further analyze the effect of noise reduction, the experiment selected four indicators of kurtosis value, cross-correlation coefficient, root mean square error (RMSE), and mean absolute error (MAE) as evaluation indicators. After calculation, the results obtained are shown in Table 5. In Table 5, by using the method proposed in this paper, the correlation value and kurtosis value obtained after noise reduction were the largest, while the RMSE and the MAE were the smallest. However, the four groups of evaluation index values obtained after noise reduction using LMD–RobustICA, EMD–RobustICA, and EEMD–RobustICA were relatively poor. Therefore, after quantitative analysis, it can be known that the signal obtained after noise reduction using the method proposed in this paper contained a higher impact component, a greater degree of correlation with the original signal, and a relatively higher waveform similarity.

Table 5. Comparison of noise-reduction effect index values.

Evaluation Index	LMD–RobustICA	EMD–RobustICA	EEMD–RobustICA	Proposed Method
Kurtosis	4.231	4.609	4.614	6.162
correlation coefficient	0.462	0.505	0.518	0.681
RMSE	1.009	0.867	0.858	0.740
MAE	0.611	0.691	0.690	0.583

Next, Hilbert envelope demodulation was performed on the results obtained based on the above three methods, and then the corresponding fault features were extracted. The envelope spectrum is shown in Figures 16–19. It can be seen from Figure 16 that the envelope spectrum obtained by the method proposed in this paper can clearly show multiple peaks with higher amplitudes, and the above peaks correspond to the frequency of one to eight times of the fault frequency, and so on. It shows that the useful signal and noise have been separated well, and the fault feature can be successfully extracted. At the same time, it can be seen from Figures 17–19 that the envelope spectrum obtained by the LMD–RobustICA, EMD–RobustICA, and EEMD–RobustICA can extracted components such as two to eight times of the fault frequency. However, it is difficult to clearly distinguish the fundamental frequency of fault frequency from each peak. In addition, compared with Figure 16, the amplitude of the fault frequency in Figures 17–19 was relatively low. It can be seen that the effect of using the method proposed in this paper to extract fault features was more significant.

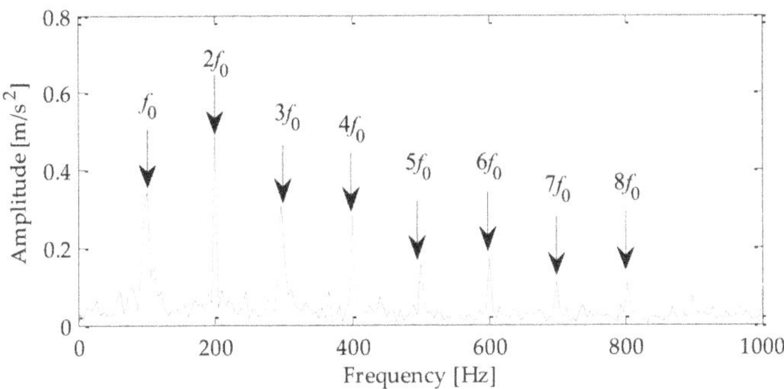

Figure 16. Analysis of signal envelope spectrum after noise reduction based on the proposed method.

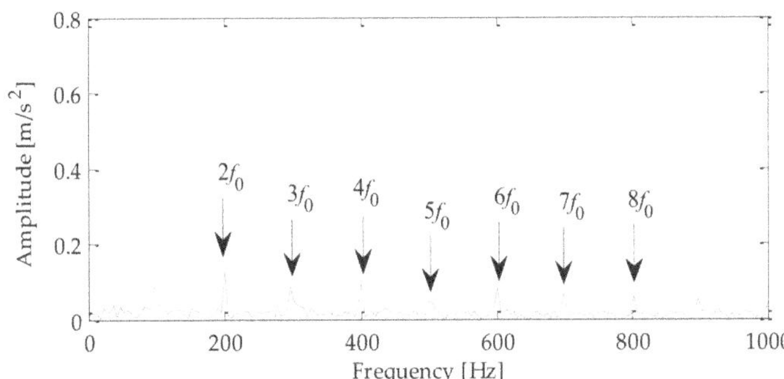

Figure 17. Analysis of signal envelope spectrum after noise reduction based on LMD–RobustICA.

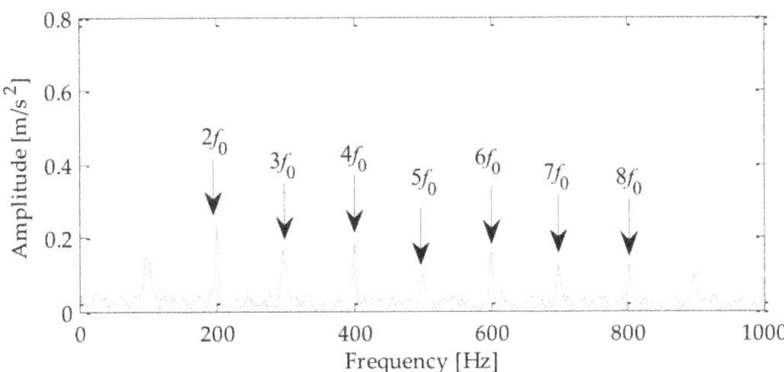

Figure 18. Analysis of signal envelope spectrum after noise reduction based on EMD–RobustICA.

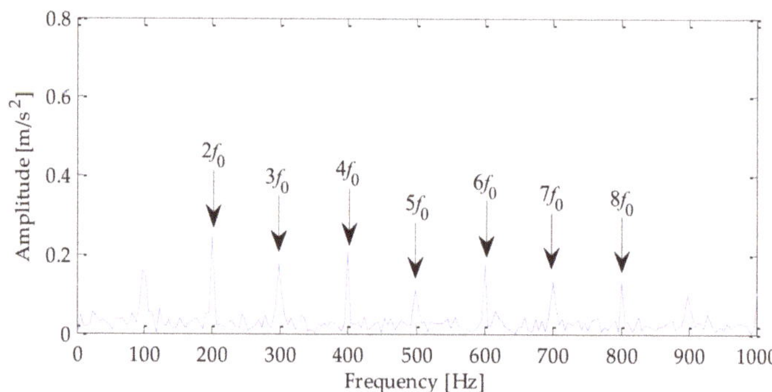

Figure 19. Analysis of signal envelope spectrum after noise reduction based on EEMD–RobustICA.

5. Case Analysis

To test whether the method proposed in this paper can effectively extract the fault characteristics of rolling bearing, an example was given to verify it. The experimental data used in the experiment were from Case Western Reserve University [33]. The structure diagram of the rolling bearing test platform and rolling bearing is shown in Figure 20 [31]. The experimental platform mainly consists of the drive motor, torque speed sensor, and power meter. Among them, the model of rolling bearing at the driving end was 6205-2RSJEMSKF. Its specific parameters are shown in Table 6. During the experiment, the set sampling frequency was 12 KHz. The acceleration sensor collected the vibration signals of the inner and outer rings with a damage diameter of 0.007 inches for experimental analysis. By substituting the relevant parameters in Table 6 into the calculation formula of fault characteristic fundamental frequency of inner ring and outer ring of rolling bearing, it can be calculated that the fundamental frequency of inner ring fault was 162.19 Hz and that of outer ring fault was 107.36 Hz.

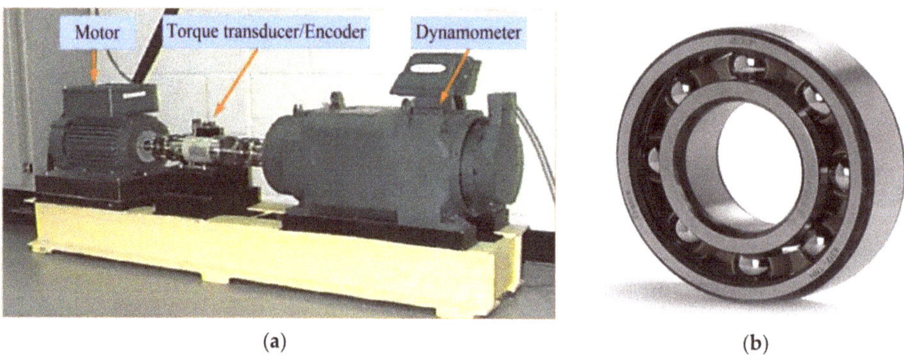

(a)

(b)

Figure 20. Illustration of the bearing experimental platform and rolling bearing. (**a**)Illustration of the bearing experimental platform. (**b**) Rolling bearing.

Table 6. The bearing structure factor.

Inner Diameter (Inches)	Outer Diameter (Inches)	Contact Angle (α)	Pitch Circle Diameter D (Inches)	Speed (rpm)
0.9843	2.0472	0^0	1.5327	1797

5.1. Inner Ring Signal Analysis

By processing the collected signal, the time domain waveform and frequency domain waveform of the fault signal of the inner ring of the rolling bearing are shown in Figures 21 and 22 can be obtained. It can be seen from that the noise interference in the signal was relatively small, and the fault impact characteristics were prominent. To test the method's effectiveness, SNR = −2 dB white Gaussian noise was added to the inner ring fault signal in this experiment. The time–domain waveform and frequency–domain waveform of the finally mixed signal is shown in Figures 23 and 24, respectively.

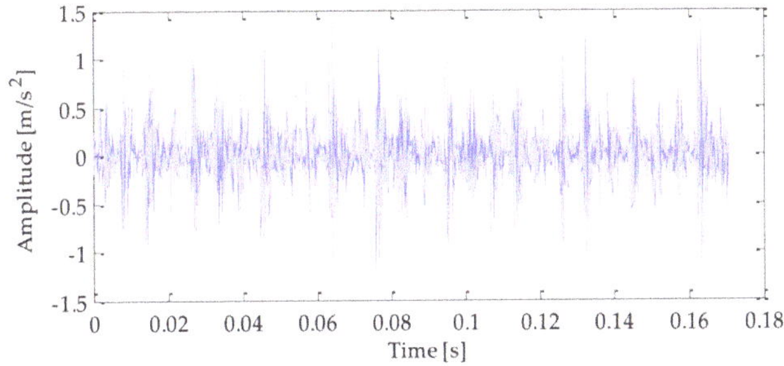

Figure 21. Time–domain analysis of the inner ring fault signal.

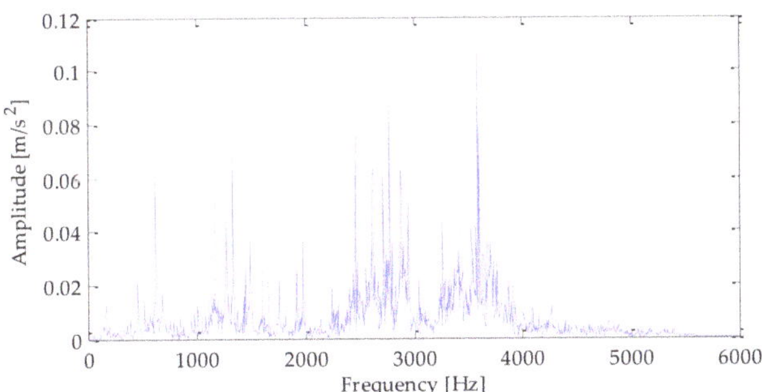

Figure 22. Frequency–domain analysis of the inner ring fault signal.

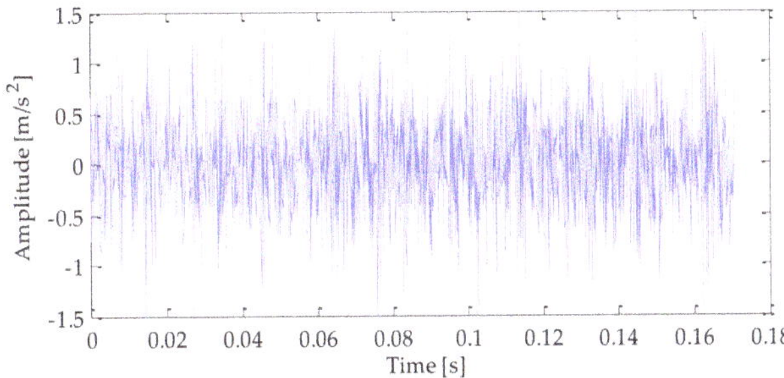

Figure 23. Time domain analysis of the mixed signal.

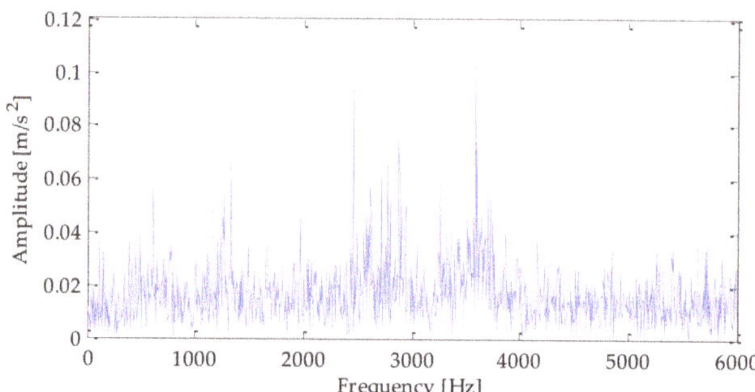

Figure 24. Frequency domain analysis of the mixed signal.

Next, the fault signal of the inner ring of the rolling bearing after adding noise will be processed. Firstly, the VMD was optimized by information entropy to select the values of parameters α and k to be initialized in VMD. The default value of α was 2500. Then, we initialized $k = 3$, and the search range of k was set to [3,15]. The value of optimal mode k was searched according to the principle of minimum envelope spectral entropy. The relationship between k and envelope spectral entropy as shown in Figure 25 can be obtained through calculation. As can be seen from Figure 25, with the continuous increase of k value, the value of the corresponding envelope spectral entropy was first decreased, then gradually increased, and, finally, decreased. Therefore, $k = 4$ was taken as the optimal value.

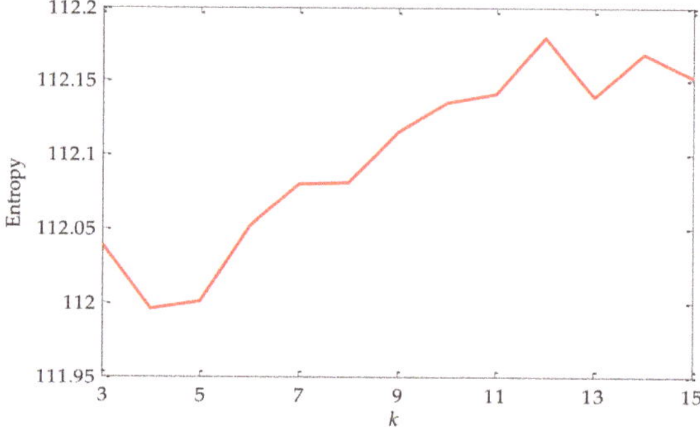

Figure 25. Curve of fitness varying with K value.

Next, the value of α was determined by experiment. The search range of α was set as [100, 2500], and then the value of the optimal mode was searched according to the principle of minimum envelope spectral entropy. The relationship between α and envelope spectral entropy as shown in Figure 26 can be obtained through calculation. As the value of α increases, the corresponding envelope spectral entropy decreases and gradually tends to be flat. When $\alpha = 2050$, the envelope spectrum entropy was the smallest. Therefore, the value of α was 2050. After the above parameter optimization, it can be obtained that the value of the optimal parameter combination k and α was [4, 2050].

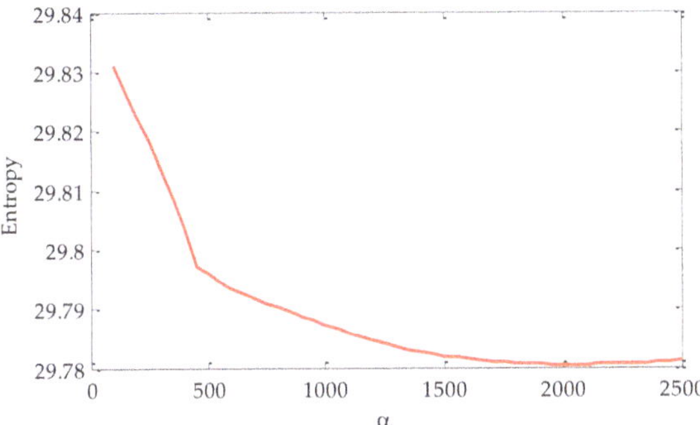

Figure 26. Curve of fitness varying with α value.

Next, this experiment substituted the selected initialization parameters into VMD, and then decomposed the signal. As is shown in Figure 27, after VMD decomposition optimized by information entropy, four IMF components were obtained. Because VMD has excellent advantages in suppressing mode aliasing and endpoint effect, it has better signal characterization ability. Next, all signal components $C(t)$ and $Q(t)$ were calculated. The calculated results are shown in Table 7. It can be seen that the IMF1 component and IMF1 component obtained by VMD decomposition met the conditions for screening the optimal signal component. That is, $C(t)$ was greater than 0.3, and $Q(t)$ was more significant than 3. Therefore, the above two IMF components were selected to reconstruct the observation signal channel. Finally, on this basis, RobustICA algorithm was used to separate signal and noise. The signal noise-reduction results obtained by using the proposed method are shown in Figure 28. Some impact characteristics of the signal can already be shown in Figure 28.

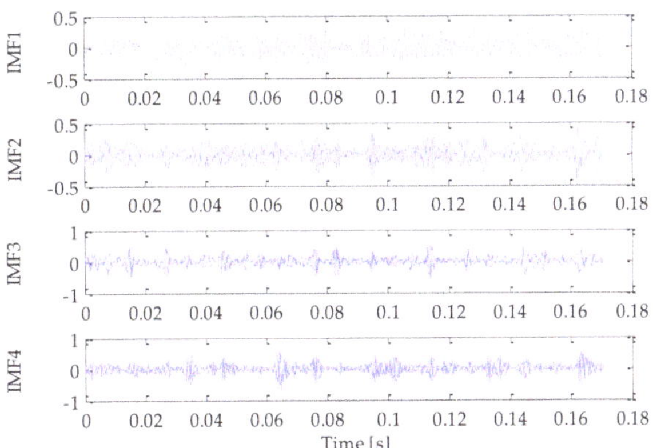

Figure 27. VMD decomposition result.

Table 7. The correlation coefficient and kurtosis between IMFand original signal (VMD).

Parameter	IMF1	IMF2	IMF3	IMF4
$C(t)$	0.354	0.396	0.512	0.524
$Q(t)$	2.661	3.044	4.174	4.082

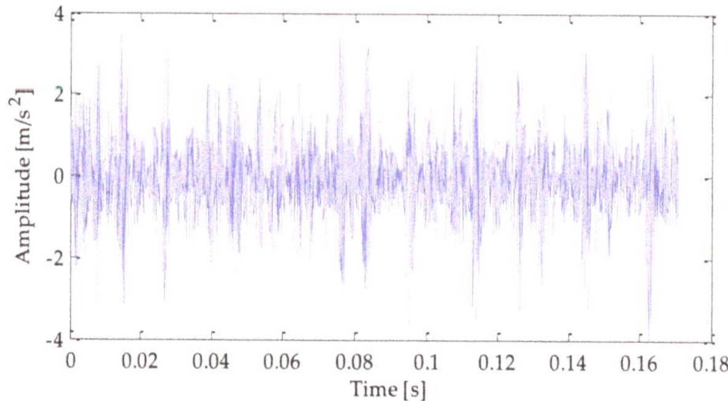

Figure 28. Noise-reduction results by the proposed method.

Finally, the denoised signal was demodulated by the Hilbert envelope, and then the corresponding fault features were extracted. To compare the experimental effects of different methods, LMD–RobustICA, EMD–RobustICA, and EEMD–RobustICA were used to denoise the signal. Then Hilbert envelope spectrum was generated.The envelope spectra obtained based on the above three methods are shown in Figures 29–32, respectively. After analyzing the above four graphs, it can be seen that from the envelope spectrum of the above methods, the one-time frequency of the fault frequency, as well as the two-time frequency and five-time frequency of the fault frequency could be extracted. However, the amplitude of the fault frequency in the envelope spectrum obtained based on the method proposed in this paper was relatively high, especially the fundamental frequency amplitude. Therefore, the effect of using the proposed method to extract fault features was more significant.

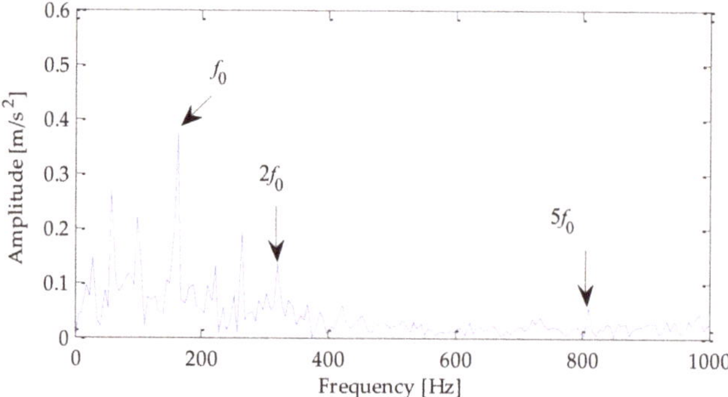

Figure 29. Analysis of signal envelope spectrum after noise reduction based on the proposed method.

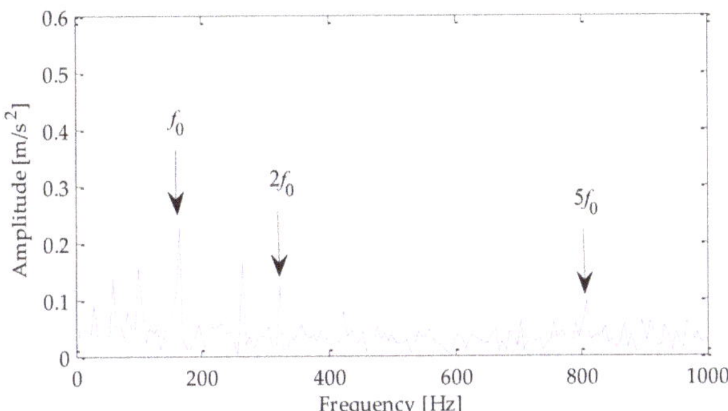

Figure 30. Analysis of signal envelope spectrum after noise reduction based on LMD–RobustICA.

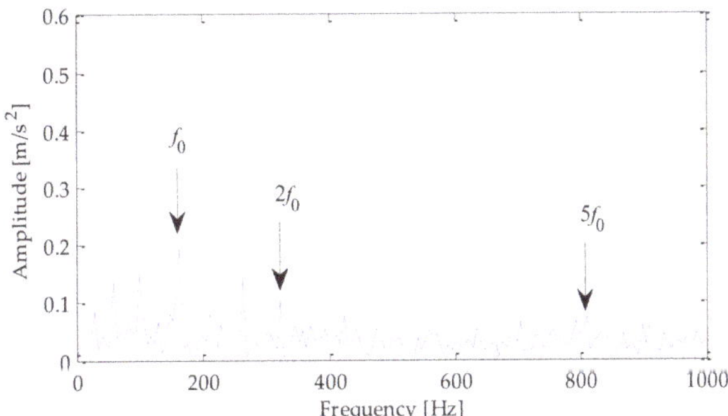

Figure 31. Analysis of signal envelope spectrum after noise reduction based on EMD–RobustICA.

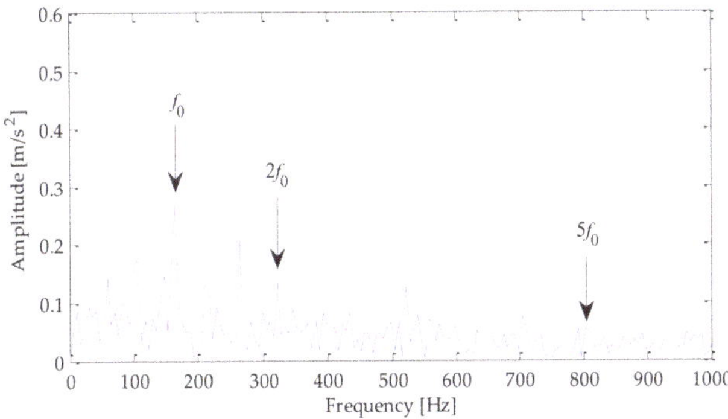

Figure 32. Analysis of signal envelope spectrum after noise reduction based on EEMD–RobustICA.

5.2. Outer Ring Signal Analysis

By extracting the rolling bearing outer ring fault signal data, the time domain waveform and frequency domain waveform of the outer ring fault signal is shown in Figures 33 and 34. Similarly, SNR = −2 dB white Gaussian noise was added to the inner ring fault signal in this experiment to test the method's effectiveness. The time domain waveform and frequency domain waveform of the finally obtained mixed signal are shown in Figures 35 and 36. Due to noise interference, it wasnot easy to distinguish the impact features from the diagram. Next, the signal was further processed.

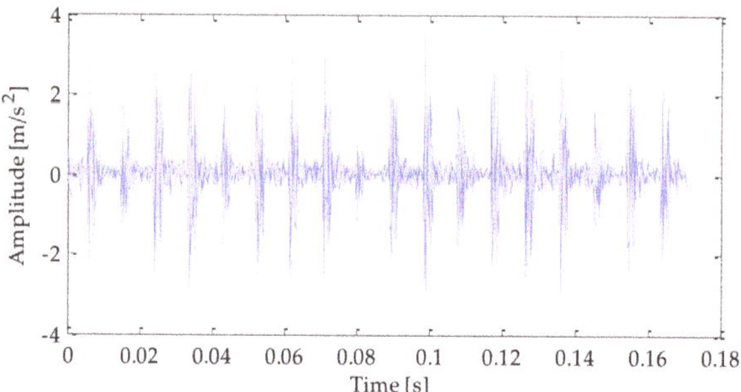

Figure 33. Time domain analysis of the outer ring fault signal.

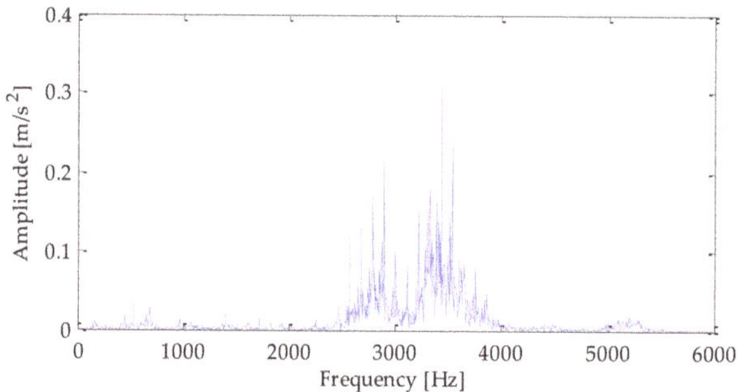

Figure 34. Frequency domain analysis of the outer ring fault signal.

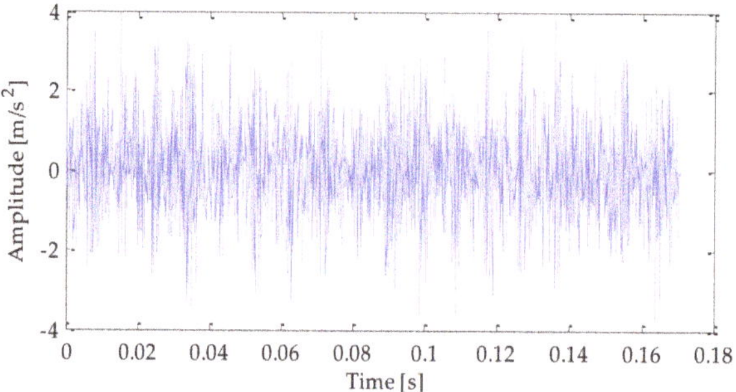

Figure 35. Time domain analysis of the mixed signal.

Next, the VMD was optimized by information entropy to select the parameters α and k was initialized. Similarly, the value of α was the default value of 2500. First, $k = 3$ was initialized, and the search range of k was set to [3,15]. The value of optimal mode kwas searched according to the principle of minimum envelope spectral entropy. Through the search, the results shown in Figure 37 can be obtained. It is shown that the corresponding envelope spectral entropy's value increased with the continuous increase of the k value. Therefore, $k = 3$ was taken as the optimal value.

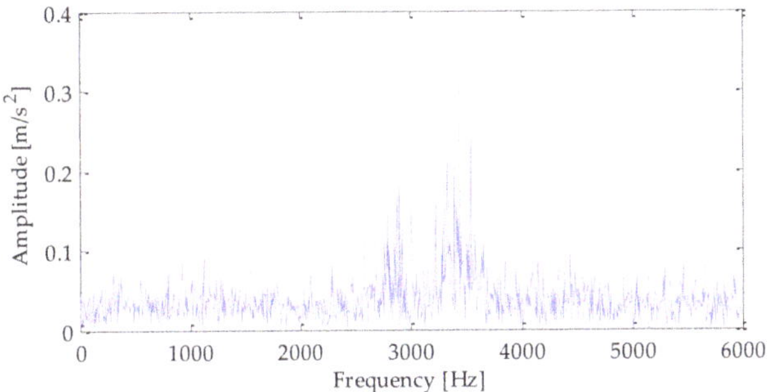

Figure 36. Frequency domain analysis of the mixed signal.

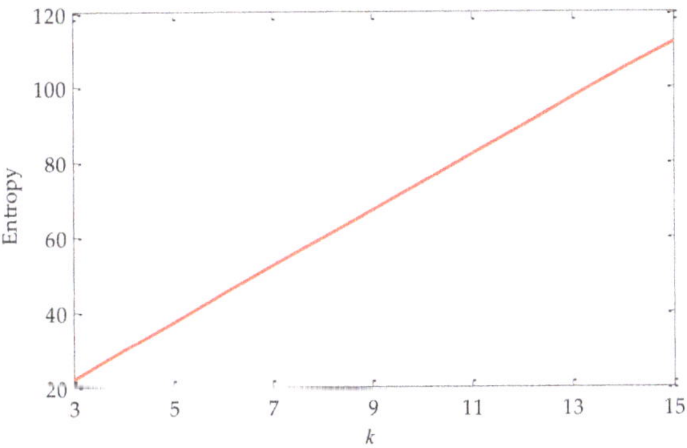

Figure 37. Curve of fitness varying with *k* value.

Based on the above calculation results, the value of α was searched. The search range of α was set as [100, 2500], and then the value of the optimal mode α was searchedaccording to the principle of minimum envelope spectral entropy. Through calculation, the relationship between α and envelope spectral entropy as shown in Figure 38 can be obtained. As shown in Figure 38, with the increasing value of α, the corresponding envelope spectral entropy was decreasing and gradually tends to be flat. When α = 1900, the envelope spectral entropy was the smallest. Therefore, the value of α was 2050. After the above parameter optimization, it can be obtained that the value of the optimal parameter combination k and α was [3, 1900].

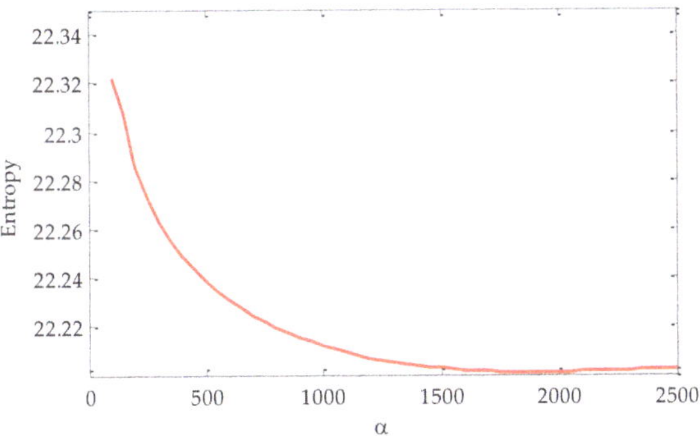

Figure 38. Curve of fitness varying with a value.

Next, the optimal parameters obtained by the search were substituted into the VMD, and then the signal was decomposed. As is shown in Figure 39, after VMD decomposition optimized by information entropy, three IMFs were obtained. Based on the above decomposition results, the $C(t)$ and $Q(t)$ of IMF1, IMF2, and IMF3 are further calculated. The calculated results are shown in Table 8. It can be seen that the IMF2 component and IMF3 component obtained by VMD decomposition met the conditions for screening the optimal signal component. That is, $C(t)$ was greater than 0.3 and $Q(t)$ was more significant than 3. Therefore, the above two signal components were selected and used to reconstruct the observation signal channel. Then, RobustICA was used for signal noise reduction. The final signal noise-reduction result is shown in Figure 40. It can be seen that the periodic impact had a high similarity with the waveform of the original signal.

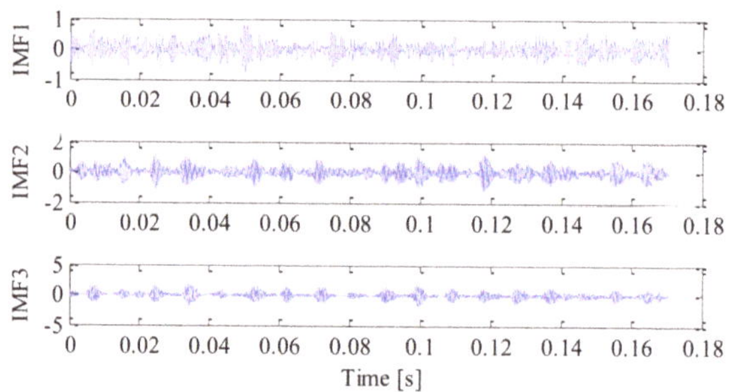

Figure 39. VMD decomposition result.

Table 8. The correlation coefficient and kurtosis between IMFand original signal (VMD).

Parameter	IMF1	IMF2	IMF3
$C(t)$	0.363	0.489	0.623
$Q(t)$	2.894	3.373	4.323

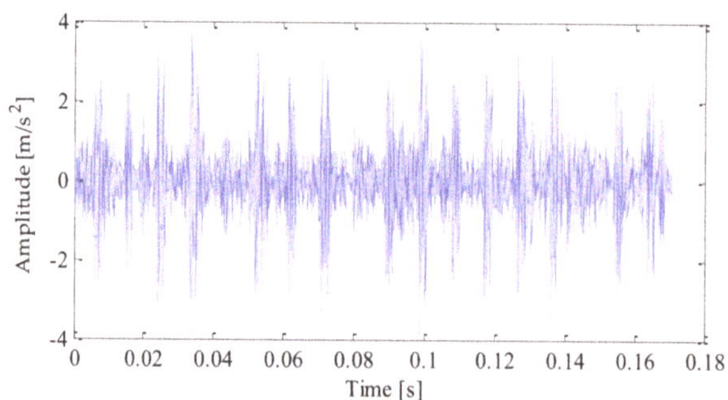

Figure 40. Noise-reduction results by the proposed method.

Next, the outer ring fault signal after signal noise reduction was demodulated by the Hilbert envelope to extract the fault feature. To furtheranalyze the fault feature extraction effect of the proposed method more intuitively, LMD–RobustICA, EMD–RobustICA, and EMD–RobustICA methods were used to denoise the signal. Then Hilbert envelope spectrum was generated. The envelope spectra obtained based on the above methods are shown in Figures 41–44, respectively. Through the analysis of Figures 41–44, it can be seen that the frequency doubling component of the outer ring fault characteristic frequency could be extracted by using the above methods for fault feature extraction. Meanwhile, the peak value of the components of the fault frequency from one to six times frequency is much higher than that obtained by the other three methods. The surrounding interference could not affect the identification of frequency doubling. However, the peak value of fault characteristic frequency in the envelope spectrum based on the other three methods was relatively low, and the interference components near these frequencies were close to the fault characteristic frequency. This brought some interference to fault feature extraction.

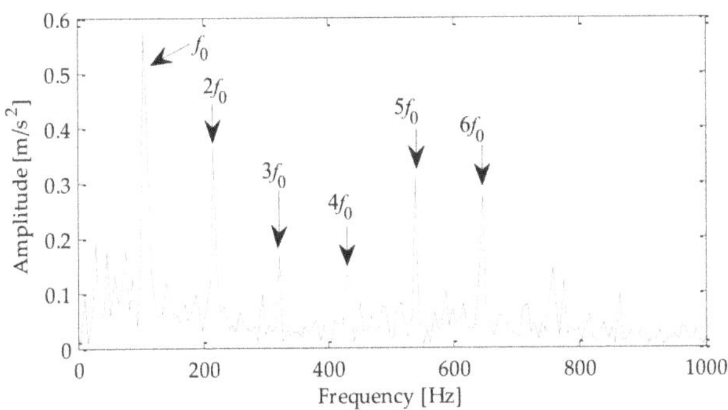

Figure 41. Analysis of signal envelope spectrum after noise reduction based on the proposed method.

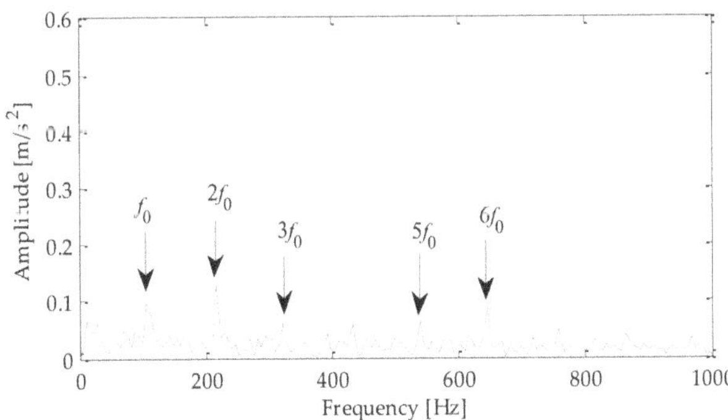

Figure 42. Analysis of signal envelope spectrum after noise reduction based on LMD–RobustICA.

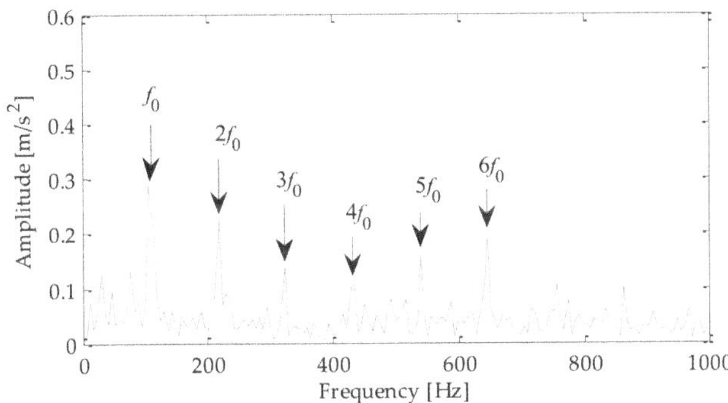

Figure 43. Analysis of signal envelope spectrum after noise reduction based on EMD–RobustICA.

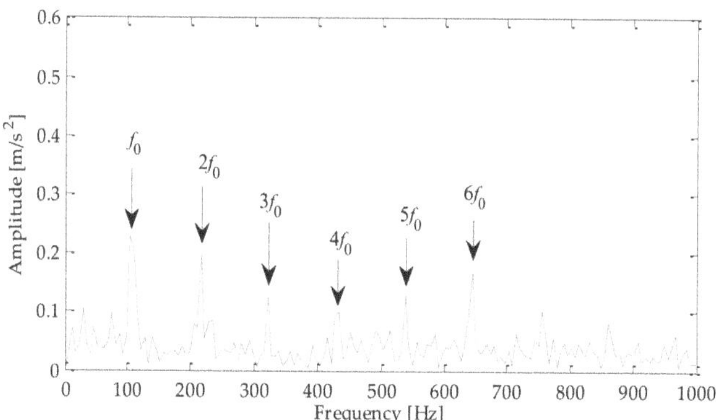

Figure 44. Analysis of signal envelope spectrum after noise reduction based on EEMD–RobustICA.

6. Conclusions

In this paper, a fault feature extraction of the rolling bearing signal under strong noise background is studied by using the combination of VMD optimized with information entropy and RobustICA. The conclusions were as follows.

(1) Although VMD can analyze the signal in the frequency domain, the effect is limited by the impact of modal component k and penalty factor α. This study used information entropy to optimize VMD to set initialization parameters. Compared with the way of setting parameters by experience, this method can search for a better combination of VMD parameters. This method can overcome the problems of modal aliasing and endpoint effect caused by impact component and noise interference in traditional EMD, LMD and EEMD, and has a good processing effect on the extraction of fault characteristic frequency of non-stationary and nonlinear signals. It can extract fault features more accurately. Compared with the traditional method, the experimental results show that this method can highlight the fault characteristic frequency and distinguish the fault.

(2) In this experiment, a typical simulation signal model is selected and Gaussian white noise is added on this basis to simulate the periodic impact signal caused by bearing fault under the condition of noise interference. Then, a signal component screening criterion based on correlation coefficient and kurtosis is established, and the optimal signal component is used to construct the observation signal channel of RobustICAalgorithm, so as to achieve the purpose of noise reduction.Through the in-depth analysis of the constructed simulation signal and the collected signal of the actual rolling bearing, it can be seen that compared with the traditional methods based on LMD–RobustICA, EMD–RobustICA, and EEMD–RobustICA, the method proposed in this paper can obtain better evaluation results of noise-reduction index, and the time–domain waveform of the signal after noise reduction is very similar to the waveform of the original signal.

(3) By comparing and analyzing the envelope demodulation results obtained by different methods, it can be seen that after the envelope spectrum analysis using the method proposed in this paper, the amplitude of fault characteristic frequency has been enhanced, and the surrounding interference will not affect the identification of fault fundamental frequency and frequency doubling, which is more convenient for fault diagnosis and analysis.

As an effective adaptive signal processing method, VMD has achieved good results in the field of fault diagnosis. However, the relevant parameters of this method need to be set in advance. In the process of parameter optimization, there is no theoretical basis for the definition of parameter search range. Therefore, in the next work, we will conduct in-depth research and further improve the parameter optimization method of VMD method.

Author Contributions: Conceptualization, J.Y.; Data curation, J.Y. and C.Z.; Formal analysis, C.Z. and X.L.; Funding acquisition, J.Y.; Methodology, J.Y.; Resources, J.Y.; Writing—original draft, J.Y.; Writing—review and editing, J.Y., C.Z., and X.L. All authors have read and agreed to the published version of the manuscript.

Funding: This work was funded by the Scientific research fund project of Baoshan University (Grant No. ZKMS202101), Special Basic Cooperative Research Programs of Yunnan Provincial Undergraduate Universities' Association (Grant No. 202001BA070001-109), collaborative education project of industry university cooperation of the Ministry of Education (Grant No. 202102049026), 10th batches of Baoshan young and middle-aged leaders training project in academic and technical (Grant No. 202109), and the Ph.D. research startup foundation of Yunnan Normal University (Grant No.01000205020503131).

Institutional Review Board Statement: Not applicable.

Informed Consent Statement: Not applicable.

Data Availability Statement: The data used to support the findings of this study are available from the corresponding author upon request.

Conflicts of Interest: The authors declare that they have no conflict of interest.

References

1. Griffin, D.W.; Lim, J.S. Signal estimation from modified short-time Fourier transform. *IEEE Trans. Acoust. Speech Signal Process.* **1984**, *32*, 236–243. [CrossRef]
2. Kwok, H.K.; Jones, D.L. Improved instantaneous frequency estimation using an adaptive short-time Fourier transform. *IEEE Trans. Signal Process.* **2015**, *48*, 2964–2972. [CrossRef]
3. Stockwell, R.G.; Mansinha, L.; Lowe, R.P. Localization of the complex spectrum: The S transform. *IEEE Trans. Signal Process.* **1996**, *44*, 998–1001. [CrossRef]
4. Boashash, B.; Black, P.J. An efficient real-time implementation of the Wigner-Ville distribution. *IEEE Trans. Acoust. Speech Signal Process.* **1987**, *35*, 1611–1618. [CrossRef]
5. Morlet, J.; Arens, G.; Fourgeau, E.; Glard, D. Wave propagation and sampling theory 1: Complex signal and scattering in multilayered media. *Geophysics* **1982**, *47*, 203–221. [CrossRef]
6. Hazarika, N.; Chen, J.Z.; Tsoi, A.C.; Sergejew, A.A. Wavelet Transform. *Signal Process.* **1997**, *59*, 61–72. [CrossRef]
7. Bonizzi, P.; Karel, J.M.; Meste, O.; Peeters, R.L.M. Singular spectrum decomposition: A new method for time series decomposition. *Adv. Adapt. Data Anal.* **2014**, *6*, 1450011. [CrossRef]
8. Saidi, L.; Ali, J.B.; Bechhoefer, E.; Benbouzid, M. Wind turbine high-speed shaft bearings health prognosis through a spectral kurtosis-derived indices and SVR. *Appl. Acoust.* **2017**, *20*, 1–8. [CrossRef]
9. Meng, L.; Xiang, J.; Zhong, Y.; Song, W. Fault diagnosis of rolling bearing based on second generation wavelet denoising and morphological filter. *J. Mech. Sci. Technol.* **2015**, *29*, 3121–3129. [CrossRef]
10. Muruganatham, B.; Sanjith, M.; Krishnakumar, B.; Murty, S.A.V.S. Roller element bearing fault diagnosis using singular spectrum analysis. *Mech. Syst. Signal Process.* **2013**, *35*, 150–166. [CrossRef]
11. Huang, N.E.; Shen, Z.; Long, S.R.; Wu, M.C.; Shih, H.H.; Zheng, Q.; Yen, N.-C.; Tung, C.C.; Liu, H.H. The empirical mode decomposition and the Hilbert spectrum for nonlinear and non-stationary time series analysis. *Proc. Math. Phys. Eng. Sci.* **1998**, *454*, 903–995. [CrossRef]
12. Yeh, J.R.; Shieh, J.S.; Huang, N.E. Ensemble empirical mode decomposition: A noise-assisted data analysis method. *Adv. Adapt. Data Anal.* **2010**, *2*, 135–156. [CrossRef]
13. Smith, J.S. The local mean decomposition and its application to EEG perception data. *J. R. Soc. Interface* **2005**, *2*, 443–454. [CrossRef] [PubMed]
14. Jiang, Z.; Wu, Y.; Li, J.; Liu, Y.; Wang, J.; Xu, X. Application of EMD and 1.5-dimensional spectrum in fault feature extraction of rolling bearing. *J. Eng.* **2019**, *2019*, 8843–8847. [CrossRef]
15. Fan, H.; Shao, S.; Zhang, X.; Wan, X.; Cao, X.; Ma, H. Intelligent Fault Diagnosis of Rolling Bearing Using FCM Clustering of EMD-PWVD Vibration Images. *IEEE Access* **2020**, *8*, 145194–145206. [CrossRef]
16. Hou, J.; Wu, Y.; Gong, H.; Ahmad, A.S.; Liu, L. A Novel Intelligent Method for Bearing Fault Diagnosis Based on EEMD Permutation Entropy and GG Clustering. *Appl. Sci.* **2020**, *10*, 386. [CrossRef]
17. Liang, J.; Wang, L.; Wu, J.; Liu, Z. Elimination of End effects in LMD Based on LSTM Network and Applications for Rolling Bearing Fault Feature Extraction. *Math. Probl. Eng.* **2020**, *2020*, 7293454. [CrossRef]
18. Dragomiretskiy, K.; Zosso, D. Variational Mode Decomposition. *IEEE Trans. Signal Process.* **2014**, *62*, 531–544. [CrossRef]
19. Ye, M.; Yan, X.; Jia, M. Rolling Bearing Fault Diagnosis Based on VMD-MPE and PSO-SVM. *Entropy* **2021**, *23*, 762. [CrossRef]
20. Li, X.; Ma, Z.; Kang, D.; Li, X. Fault diagnosis for rolling bearing based on VMD-FRFT. *Measurement* **2020**, *155*, 107554. [CrossRef]

21. Xing, T.T.; Zeng, Y.; Meng, Z.; Guo, X. A fault diagnosis method of rolling bearing based on VMD Tsallis entropy and FCM clustering. *Multimed. Tools Appl.* **2020**, *79*, 30069–30085.
22. Albataineh, Z.; Salem, F.M. ARobustICA-based algorithmic system for blind separation of convolutive mixtures. *Int. J. Speech Technol.* **2021**, *24*, 701–713. [CrossRef]
23. Li, Q.; Li, M.; Zhang, X.; Ba, W. Research on Mixed Signal Separation method for Pipeline Leakage Based on RobustICA. In Proceedings of the International Conference on Robotics & Automation Engineering, Jeju, Korea, 27–29 August 2016; IEEE: Washington, DC, USA, 2016; pp. 70–74.
24. Yao, J.; Xiang, Y.; Qian, S.; Zhang, G. Cylinder Head Vibration Signals Separation of Internal Combustion Engines Based on VMD and RobustICA. *China Mech. Eng.* **2018**, *29*, 923–929.
25. Yang, J.Z.; Wang, X.D.; Feng, Z. A Noise Reduction Method Based on CEEMD-RobustICA and Its application in Pipeline Blockage Signal. *J. Appl. Sci. Eng.* **2019**, *22*, 1–10.
26. Yao, J.; Xiang, Y.; Qian, S.; Wang, S. Radiation Noise Separation of Internal Combustion Engine Based on Gammatone-RobustICA Method. *Shock. Vib.* **2017**, *2017*, 7565041. [CrossRef]
27. Zhao, F.; Wu, M. Application of EEMD- RobustICA and prony algorithm in modes identification of power system low frequency oscillation. *Acta Energ. Sol. Sin.* **2019**, *40*, 2919–2929.
28. Shannon, C.E. A mathematical theory of communication. *Bell Syst. Tech. J.* **1948**, *27*, 379–423. [CrossRef]
29. Zarzoso, V.; Comon, P. Robust independent component analysis for blind source separation and extraction with application in electrocardiography. In Proceedings of the 30th Annual International Conference of the IEEE Engineering in Medicine and Biology Society, Vancouver, BC, Canada, 20–25 August 2008.
30. Zarzoso, V.; Comon, P. Robust independent component analysis by iterative maximization of the kurtosis contrast with algebraic optimal step size. *IEEE Trans. Neural Netw.* **2010**, *21*, 248–261. [CrossRef]
31. Yang, J.Z.; Li, X.F.; Wu, L.M. Research on Fault Feature Extraction Method Based on FDM-RobustICA and MOMEDA. *Math. Probl. Eng.* **2020**, *2020*, 6753949. [CrossRef]
32. Feng, Z.; Ma, J.; Wang, X.; Wu, J.; Zhou, C. An Optimal Resonant Frequency Band Feature Extraction Method Based on Empirical Wavelet Transform. *Entropy* **2019**, *21*, 135. [CrossRef]
33. Smith, W.A.; Randall, R.B. Rolling element bearing diagnostics using the Case Western Reserve University date: A benchmark study. *Mech. Syst. Signal Process. Method* **2015**, *64*, 100–131. [CrossRef]

Article

A Fault Feature Extraction Method Based on LMD and Wavelet Packet Denoising

Jingzong Yang [1] and Chengjiang Zhou [2,*]

1   School of Big Data, Baoshan University, Baoshan 678000, China; yjingzong@foxmail.com
2   School of Information Science and Technology, Yunnan Normal University, Kunming 650500, China
*   Correspondence: chengjiangzhou@foxmail.com

Abstract: Aiming at the problem of fault feature extraction of a diaphragm pump check valve, a fault feature extraction method based on local mean decomposition (LMD) and wavelet packet transform is proposed. Firstly, the collected vibration signal was decomposed by LMD. After several amplitude modulation (AM) and frequency modulation (FM) components were obtained, the effective components were selected according to the Kullback-Leible (K-L) divergence of all component signals for reconstruction. Then, wavelet packet transform was used to denoise the reconstructed signal. Finally, the characteristics of the fault signal were extracted by Hilbert envelope spectrum analysis. Through experimental analysis, the results show that compared with other traditional methods, the proposed method can effectively overcome the phenomenon of mode aliasing and extract the fault characteristics of a check valve more effectively. Experiments show that this method is feasible in the fault diagnosis of check valve.

Keywords: checkvalve; local mean decomposition; wavelet packet transform; K-L divergence

1. Introduction

With the vigorous development of mineral pipeline transportation technology, high-pressure diaphragm pump operation, maintenance, and fault monitoring have become a concern. The high-pressure diaphragm pump provides the power required in the process of mineral transportation [1], and most of its failures are caused by one of the core components of the pump check valve. The check valve is a directional element used to control the feeding and discharging, so that the conveying medium flows in one direction and cannot flow back. Generally, the average price of each diaphragm pump exceeds CNY20 million, and the daily delivery of pulp exceeds 30,000 tons. If the stroke coefficient of the high-pressure diaphragm pump is 50 R/min, the feed and discharge check valves in a normal operating day need to act in a reciprocating manner 72,000 times. Therefore, the check valve is the most frequent and fault-prone component in a high-pressure diaphragm pump. Local faults and unplanned shutdowns in its operation state easily cause equipment pressure and flow fluctuations, resulting in equipment vibration and damage to the mineral transmission pipeline. To sum up, monitoring the operation state of the check valve is very important for the safe, stable, and efficient operation of mineral pipeline transportation. By detecting the vibration signal of the check valve and extracting the characteristics that can characterize the operation state for diagnosis and identification, the engineers and technicians can learn the operation state of the equipment in time and significantly improve the ability of operation and maintenance management. However, in the actual detection process, due to the complex field background noise, there is often strong impulse interference and random noise, making it difficult to distinguish the useful signal and noise interference, affecting the effective extraction of fault features.

Traditional signal processing methods are mostly based on Fourier transform, such as short-time Fourier transform, Wigner–Ville, power spectrum analysis, etc. [2–4]. Although

Citation: Yang, J.; Zhou, C. A Fault Feature Extraction Method Based on LMD and Wavelet Packet Denoising. *Coatings* **2022**, *12*, 156. https:// doi.org/ 10.3390/coatings12020156

Academic Editors: Ke Feng, Jinde Zheng and Qing Ni

Received: 23 December 2021
Accepted: 24 January 2022
Published: 27 January 2022

Publisher's Note: MDPI stays neutral with regard to jurisdictional claims in published maps and institutional affiliations.

the above methods have been popularized in many fields of noise reduction, there are still some defects: they cannot decompose the signal adaptively. The vibration signal of the check valve is a typical non-stationary and nonlinear signal. Before processing and analyzing it, it is necessary to obtain the local properties of the signal in the time domain. Through the ongoing efforts of scholars in related fields, some breakthroughs have been made. The empirical mode decomposition (EMD) proposed by Huang et al. [5] can adaptively decompose the signal into a series of intrinsic mode functions (IMF). However, this method has problems such as the endpoint effect, mode aliasing, etc. In recent years, the local mean decomposition (LMD) [6] proposed by Smith is a new time-frequency analysis method, which can adaptively decompose the signal to be processed into several product functions (PF) with high to low frequency and has been gradually applied. Compared with traditional methods, LMD has the advantages of better adaptability and time-frequency aggregation, solves the problems of over the envelope and under the envelope of EMD, has fewer iterations, and effectively improves the accuracy of signal decomposition. Dong Linlu et al. [7] combined the respective advantages of LMD and the singular-value decomposition method to denoise the noisy micro-vibration signal. The results show that the proposed method can better remove the high-frequency noise of the micro-vibration signal, which lays a foundation for further analysis of vibration signals. To solve the influence of complex noise and measuring point position in the monitoring process, Wang Haijun et al. [8] proposed a processing method combining a data fusion algorithm and LMD. The results show that the combined method is better than the single digital filtering method, and the vibration monitoring of a hydropower plant is realized better. Wang Zhijian et al. [9] introduced the mask signal method to process the component signals screened after LMD decomposition, which effectively suppressed the noise and weakened the mode aliasing phenomenon in the process of signal decomposition. Through the signal's LMD decomposition, the complete time-frequency distribution information can be obtained, and it is convenient to extract the signal characteristics. At the same time, the latest applications of the LMD method in other fields have been reported. Gupta P et al. [10] proposed the method of combining LMD and an artificial neural network to complete the tool chatter feature extraction in the turning process. From the experimental analysis, it can be seen that the obtained safe cutting zone is important and can limit chatter. Liao L et al. [11] completed the extraction of low-frequency noise under the background of aerodynamic noise by using the improved LMD method. Through the wind turbine experiment, the results show that the signal noise extraction effect is good, and the calculation efficiency of the algorithm is good. To solve the problematic identification of welding quality, Huang y et al. [12] proposed a classification algorithm combining LMD and the deep belief network (DBN). Experimental results show that this method has a higher recognition rate than traditional classification methods. Lee CY et al. [13] introduced LMD, wavelet packet decomposition (WPD), and other ways to solve the problems existing in the rotor fault diagnosis of an induction motor. The results show that the model can reduce the dimension of the original data and has high robustness. However, due to noise interference, the extracted PF component will be distorted to a certain extent, which affects the interpretation of the useful components of the signal. Using wavelet packet to process the fault signal can effectively overcome the problem of low-frequency resolution in the high-frequency band of the traditional wavelet transform [14] and accurately describe the signal information, which is conducive to the effective elimination of noise information.

However, due to the bad working environment and changeable working conditions of the high-pressure diaphragm pump, some noise will still remain in the signal component decomposed by the time-frequency analysis method. If fault feature extraction and analysis are carried out directly, it will have a certain impact on fault diagnosis results. Therefore, it is also necessary to reduce the noise of the signal. Wavelet packet transform is based on wavelet transform. Compared with wavelet transform, wavelet packet transform has higher-frequency resolution in high-frequency and low-frequency bands and has good adaptability. In recent years, wavelet transform has been widely used in the noise reduction of non-stationary signals. Kuanfang he et al. [15] proposed a wavelet packet denoising method for welding acoustic emission signals. The experimental results show that the proposed method can effectively process the acoustic emission signals of welding cracks. Wang X et al. [16] proposed a method based on optimized variational mode decomposition (VMD) and wavelet packet threshold denoising and applied it to remove strong white noise signals. The results show that this method retains the practical components of the signal well. Sun W et al. [17] used a wavelet packet to remove the noise in the collected bearing signal and then combined it with the LMD method to extract the fault feature. The results show that this method can effectively extract fault features.

Based on the above analysis, a fault feature extraction method based on local mean decomposition (LMD) and wavelet packet denoising is proposed in this paper. Firstly, the vibration signal of the check valve was decomposed by LMD, and the K-L divergence of each AM-FM component was obtained. The practical component signal was selected for reconstruction through the comparative analysis with the set threshold. Then, wavelet packet transform was used to denoise the reconstructed signal to reduce the noise interference. Finally, Hilbert envelope spectrum analysis asper formed on the denoised signal to extract the fault signal features. The effectiveness of the proposed method was verified by analyzing the actual fault data of the check valve.

The rest of the paper is organized as follows: In Section 2, we introduce the basic principle of the LMD algorithm and K-L divergence and describe the process of PF component selection. In Sections 3 and 4, we introduce the basic principles of Hilbert envelope demodulation and wavelet packet denoising. In Section 5, we describe the modeling process of check valve fault feature extraction based on LMD and wavelet packet denoising. Section 6 is experimental analysis. The effectiveness of the method proposed in this paper was verified by the collected actual fault data of the check valve. Section 7 concludes the paper.

2. LMD Algorithm Based on K-L Divergence

2.1. LMD Algorithm

LMD is a time-frequency analysis method proposed by Smith. Its principle is to decompose the signal into product function components of different frequencies and a margin, and each component is obtained by multiplying a pure FM signal and envelope signal. If the original signal is $x(t)$, the decomposition step can be described in the following form.

(1) Calculate all local extreme points of the original signal $x(t)$, and then calculate the average value m_i of adjacent local extreme points n_i and n_{i+1}:

$$m_i = \frac{n_i + n_{i+1}}{2} \tag{1}$$

The moving average method is used to smooth the line composed of m_i, and the local mean function $m_{11}(t)$ can be obtained.

(2) The envelope estimation value a_i is calculated according to the adjacent extreme points n_i and n_{i+1}:

$$a_i = \frac{|n_i - n_{i-1}|}{2} \tag{2}$$

Similarly, the envelope estimation function $a_{11}(t)$ is obtained by the moving average method.

(3) The local mean function $m_{11}(t)$ is separated from the original signal $x(t)$.

$$h_{11}(t) = x(t) - m_{11}(t) \tag{3}$$

(4) Divide $h_{11}(t)$ by $a_{11}(t)$ to obtain FM signal $s_{11}(t)$

$$s_{11}(t) = h_{11}(t)/a_{11}(t) \tag{4}$$

When $a_{12}(t) = 1$, $s_{11}(t)$ is a standard pure FM signal, and when $a_{12}(t)$ is not equal to 1, $s_{11}(t)$ is taken as the original data, and the above processes repeated. Until $a_{1(n+1)}(t) = 1$ is met, the pure FM signal $s_{1n}(t)$ is calculated, as shown in the following formula:

$$\begin{cases} h_{11}(t) = x(t) - m_{11}(t) \\ h_{12}(t) = x(t) - m_{12}(t) \\ \quad \vdots \\ h_{1n}(t) = x(t) - m_{1n}(t) \end{cases} \tag{5}$$

$$\begin{cases} s_{11}(t) = h_{11}(t)/a_{11}(t) \\ s_{12}(t) = h_{12}(t)/a_{12}(t) \\ \quad \vdots \\ s_{1n}(t) = h_{1n}(t)/a_{1n}(t) \end{cases} \tag{6}$$

The conditions for iteration termination are as follows:

$$\lim_{n \to \infty} a_{1n}(t) = 1 \tag{7}$$

(5) Calculate the envelope signal:

$$a_1(t) = a_{11}(t)a_{12}(t) \cdots a_{1n}(t) = \prod_{q=1}^{n} a_{1q}(t) \tag{8}$$

(6) The product function can be obtained by multiplying the pure FM signal $s_{1n}(t)$ and the envelope signal $a_1(t)$

$$PF_1(t) = a_1(t)s_{1n}(t) \tag{9}$$

(7) Separate $PF_1(t)$ from $x(t)$ and find the signal $u_1(t)$. Then, repeat the above steps k times. When $u_k(t)$ becomes a monotone function, the operation is terminated.

$$\begin{cases} u_1(t) = x(t) - PF_1(t) \\[2mm] u_2(t) = u_1(t) - PF_2(t) \\ \quad \vdots \\ u_k(t) = u_{k-1}(t) - PF_k(t) \end{cases} \tag{10}$$

The following formula can be obtained by combining all PF components and recombining.

$$x(t) = \sum_{p=1}^{k} PF_p(t) + u_k(t) \tag{11}$$

From the above description, it can be seen that the LMD is an adaptive decomposition method based on the local extreme points of the signal itself. The decomposition process is a multi-cycle iterative process.

2.2. PF Component Selection Based on K-L Divergence

Several PF components obtained by LMD have a different correlation with the original signal. To effectively screen component signals, appropriate screening methods must be adopted. K-L divergence is called directional divergence, describing the difference between the two probability distributions [18]. Let the two probability distributions be $p_1(x)$ and $p_2(x)$, then the K-L distance is defined as follows:

$$\delta(p_1, p_2) = \int p_1(x) \log \frac{p_1(x)}{p_2(x)} dx \tag{12}$$

The K-L divergence values of $p_1(x)$ and $p_2(x)$ are:

$$D(p_1, p_2) = \delta(p_1, p_2) + \delta(p_2, p_1) \tag{13}$$

The vibration signals collected in the equipment characterize the vibration of the equipment. Therefore, assuming that the probabilities between the two signals are $p_1(x)$ and $p_2(x)$, the closer the vibration between them, the closer $p_1(x)$ and $p_2(x)$ will be, and the smaller the K-L divergence value will be.

Since several PF components are obtained through LMD, the solution of K-L divergence for the original signal $X = \{x_1, x_2 \ldots x_n\}$ and component $Y = \{y_1, y_2 \ldots y_n\}$ decomposed by the LMD algorithm is as follows:

(1) Solve the probability distribution of the above two signals and let the function $p_1(x)$ be the kernel density estimation of signal X:

$$p_1(x) = \frac{1}{nh} k \sum_{i=1}^{n} \left[\frac{x_i - x}{h} \right], \; x \in R \tag{14}$$

where $k(\cdot)$ is the kernel function, which usually uses the Gaussian kernel function, which is $k(u) = \frac{1}{\sqrt{2\pi}} e^{-u^2/2}$, and h is a given positive number. Similarly, $p_2(x)$ can be obtained in the above way.

(2) The K-L distances $X = \{x_1, x_2 \ldots x_n\}$ and $Y = \{y_1, y_2 \ldots y_n\}$ are obtained by substituting the probability distributions of signal $\delta(p_1, p_2)$ and signal $\delta(p_2, p_1)$ into Formula (12).

(3) By substituting the calculated $\delta(p_1, p_2)$ and $\delta(p_2, p_1)$ into Formula (13), the corresponding K-L divergence can be obtained.

(4) Normalize all calculated K-L divergence values.

(5) Filter according to the preset threshold. If the component is less than the threshold, it will be regarded as the component containing an obvious fault signal. According to the data obtained in this study, the set threshold is 0.03.

3. Hilbert Envelope Demodulation

The Hilbert envelope demodulation method mainly converts the actual signal into an analytical signal by Hilbert transform and then takes the modulus to obtain its envelope. Assuming that the signal $x(t)$ has amplitude modulation envelope $A(t)$ and phase modulation function $\varphi(t)$, its expression is as follows:

$$x(t) = A(t) \cos(2\pi f t + \varphi(t)) \tag{15}$$

The Hilbert transform of signal $x(t)$ can be an approximately $90°$ phase shift of $x(t)$:

$$\hat{x}(t) = A(t) \sin(2\pi f t + \varphi(t)) \tag{16}$$

Construct the analytical signal $Z(t)$ so that:

$$Z(t) = x(t) + j\hat{x} = A(t) e^{j\varphi(t)} \tag{17}$$

By calculating the modulus of the above formula, the following formula can be obtained:

$$|Z(t)| = A(t) \tag{18}$$

In the above formula, $|Z(t)|$ is the envelope of the signal.

4. Principle of Wavelet Packet Denoising

As one of the more detailed signal analysis methods, the advantage of wavelet packet transform is that it can adaptively select the corresponding frequency band and divide the frequency band at multiple levels [19,20]. For a one-dimensional signal containing noise, it can be described as the following formula:

$$S(i) = f(i) + \alpha\beta(i)\ (i = 0, 1, 2, \ldots, n - 1) \tag{19}$$

where $f(i)$ is the actual signal, $S(i)$ is the signal containing noise, and $\beta(i)$ is noise. The process of wavelet packet denoising is as follows:

(1) Determine the corresponding wavelet base and decomposition level, and then start wavelet packet decomposition;

(2) According to the established entropy standard, the optimal tree is obtained, and the optimal wavelet basis is determined;

(3) The appropriate threshold is selected, and the high-frequency coefficients with different decomposition scales are quantified at the same time;

(4) The signal reconstruction operation is carried out according to the decomposition coefficients and quantization coefficients of the N-th layer wavelet packet.

The decomposition structure of wavelet packet is shown in Figure 1 [21].

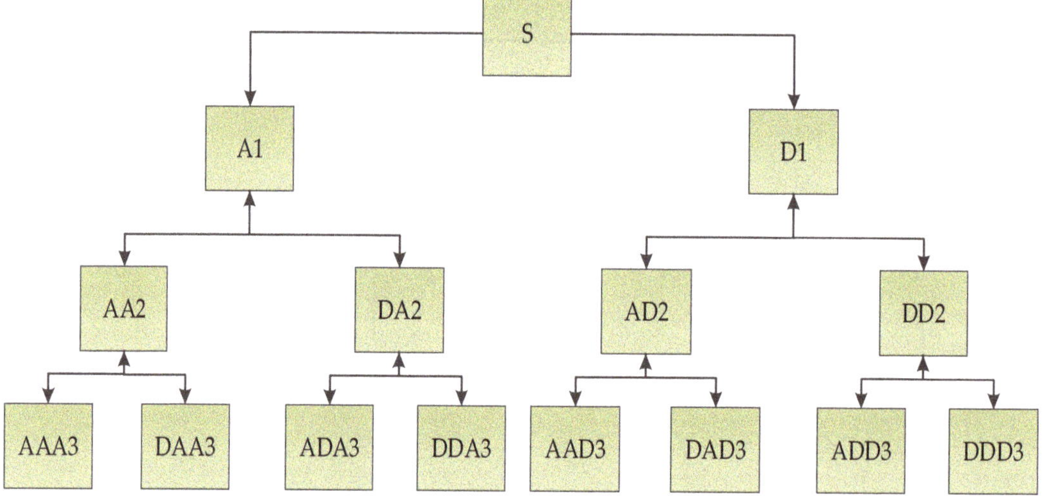

Figure 1. Decomposition structure of wavelet packet.

5. Fault Feature Extraction Model of Check Valve Based on Local Mean Decomposition Wavelet Packet Denoising

Based on the above theoretical analysis, a check valve fault feature extraction method based on local mean decomposition wavelet packet denoising is proposed in this paper. The research outline of the proposed method is shown in Figure 2, and the method implementation process is shown in Figure 3. The specific steps are as follows:

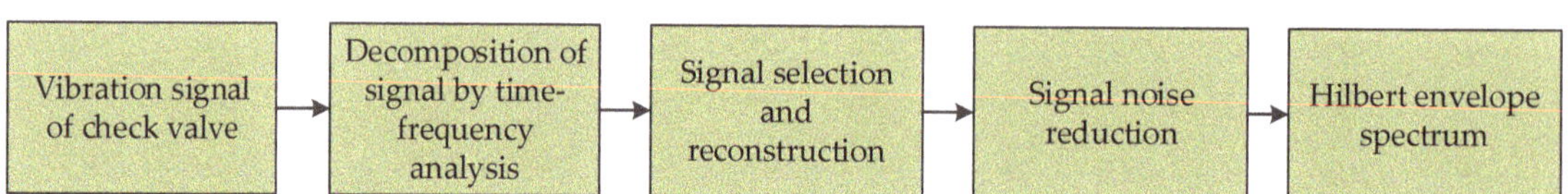

Figure 2. Research outline of the proposed method.

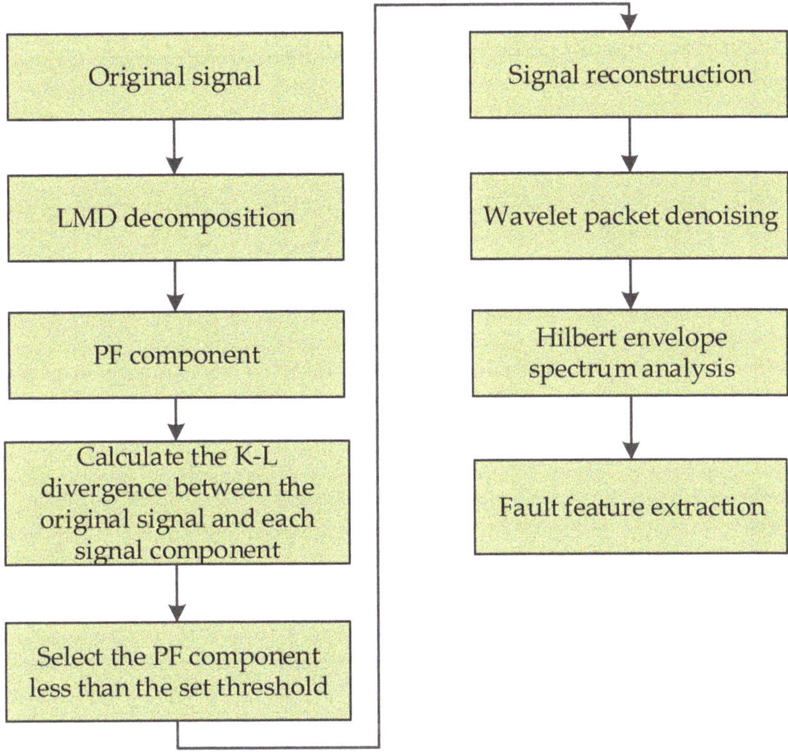

Figure 3. Fault feature extraction process of check valve based on LMD and wavelet packet denoising.

(1) The experimental failure data of wear breakdown of check valve are selected, and then the above samples are decomposed into several PF components by LMD decomposition.

(2) According to the calculation method of K-L divergence, the K-L divergence between the original signal and each PF component obtained by decomposition is calculated. Then, all K-L divergence values are normalized.

(3) According to the set threshold, the calculated K-L divergence value is compared with it, and then any PF component less than the threshold is filtered.

(4) The filtered PF components are reconstructed, and then wavelet packet denoising is carried out.

(5) Hilbert envelope demodulation is performed on the denoised signal.

6. Experimental Analysis

The field data of the high-pressure diaphragm pump check valve obtained by the data acquisition system of a slurry transmission pipeline in Western China were used in the experiment. The diaphragm pump used in the slurry transmission pipeline pump station was of the TZPM series, and the data acquisition card used was the PXIe-3342 eight-channel acquisition card. Through the acceleration sensor arranged outside the diaphragm pump, the vibration signal was collected and transmitted to the computer. Then, the collected data of the check valve in the wear fault state were analyzed to extract its fault characteristics. The Schematic diagram of high-pressure diaphragm pump and fault check valve is shown in Figure 4 [22]. The vibration signal acquisition system diagram of check valve is shown in Figure 5 [22].

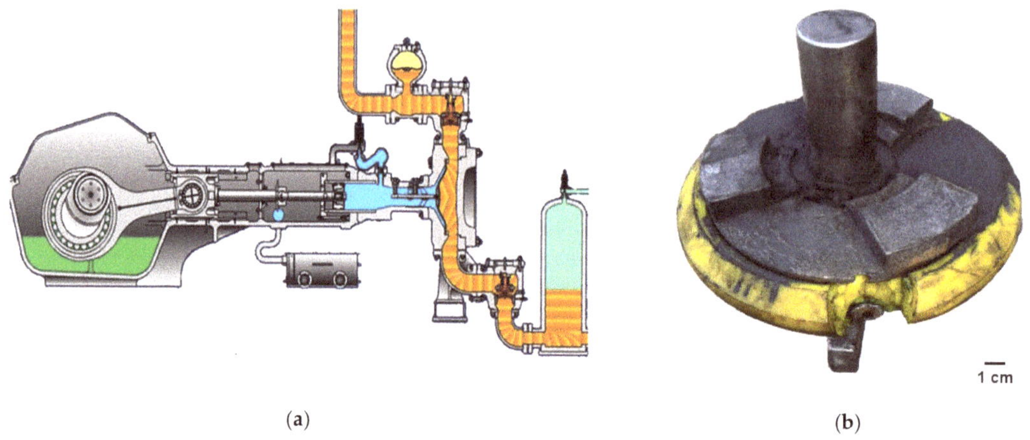

(**a**)   (**b**)

Figure 4. Schematic diagram of high-pressure diaphragm pump and fault check valve. (**a**) High pressure diaphragm pump; and (**b**) fault check valve.

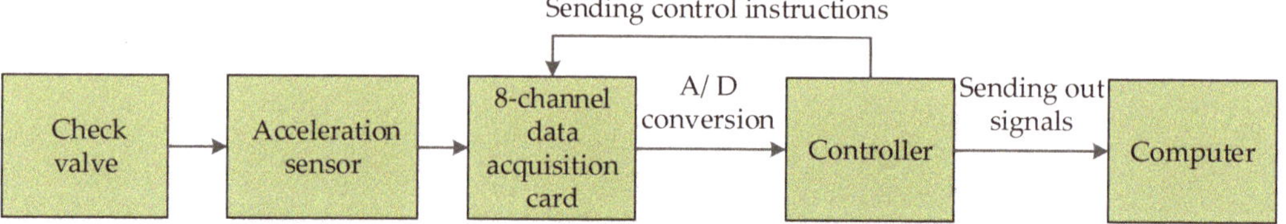

Figure 5. Vibration signal acquisition system diagram of check valve.

In the check valve, wear breakdown caused by coarse particles stuck in the valve is very common. The check valve of the high-pressure diaphragm pump has a cone-valve-type structure, and the main components are the valve body, valve core, and spring. Among them, the "valve core spring" constitutes a low-damping oscillation system, and its frequency is:

$$f = \frac{1}{2\pi}\sqrt{\frac{k}{m_s}} \tag{20}$$

where k is the stiffness value of the spring; and m_s is the equivalent mass value. Set the spring stiffness as follows according to the activity of the plunger of the pump

$$k = 4\pi^2 (2f)^2 m_s \tag{21}$$

where f is the frequency of normal operation of the high-pressure diaphragm pump. Its normal operating frequency is 0.5~0.517 Hz, and the frequency of the spring valve core system $fp = 2f$, namely, 1~1.034 Hz. The check valve will show the corresponding fault fundamental frequency and double frequency when it fails.

The check valves of the high-pressure diaphragm pump are matched by the feed valve and discharge valve in pairs. Therefore, the normal operation vibration signal and fault vibration signal of a group of feed and discharge check valves were randomly selected as a group of experimental data for analysis. The fault data of the check valve used in the experiment were the data when a wear breakdown fault occurred. The sampling frequency of this experiment was 2560 Hz, and the sampling data length was 10,240. After A/D conversion, the collected vibration acceleration signal was input into the computer through the controller.

Next, the normal operation and fault operation data of 10,240 check valves were taken for analysis. Figure 6 is the time-domain waveform diagram of the normal operation of the check valve. Figures 7 and 8 are the time-domain diagram and frequency-domain diagram when the check valve fails, respectively. Through comparative analysis of Figures 6 and 7, it can be seen that when the check valve was in normal operation, the vibration signal contains prominent impact components, and the amplitude was relatively more apparent. It can be seen from Figure 8 that the operating frequency was mainly within 200 Hz, but it was impossible to conclude whether it had a fault. To further analyze the signal, LMD decomposition was carried out first. As shown in Figure 9, the original signal was decomposed into several PF components and a residual component. To compare the decomposition effect, empirical mode decomposition (EMD) was used to decompose the signal. The results are shown in Figure 10. After EMD decomposition, more component signals were obtained, and there was a certain degree of modal aliasing. Through comparison, it can be seen that the effect of signal decomposition using the LMD method was better, as it reduced the phenomenon of mode aliasing to a great extent. Therefore, the obtained signal component contained more information. At the same time, LMD had fewer iterations. Next, we calculated the K-L divergence of all component signals decomposed by LMD, EMD, and the original signal. The results are shown in Tables 1 and 2.

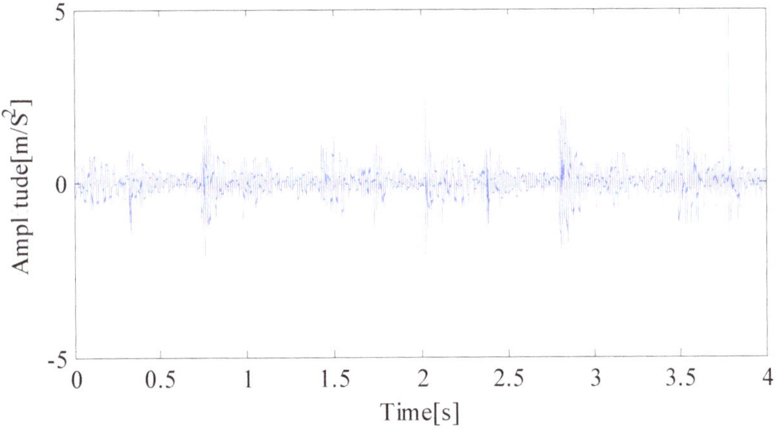

Figure 6. Time-domain waveform of check valve in normal operation.

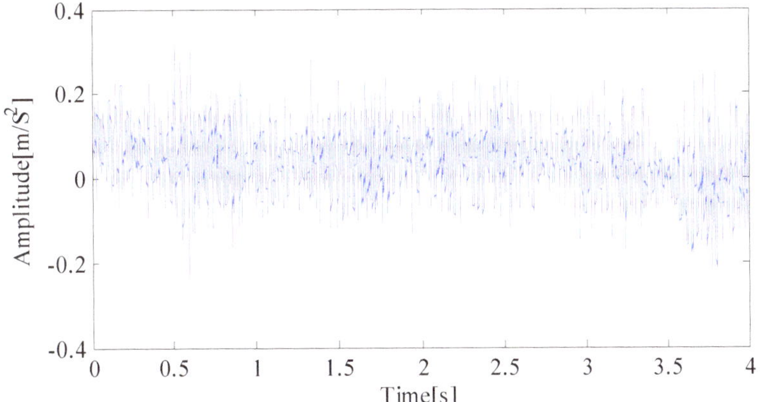

Figure 7. Time-domain waveform of check valve fault operation.

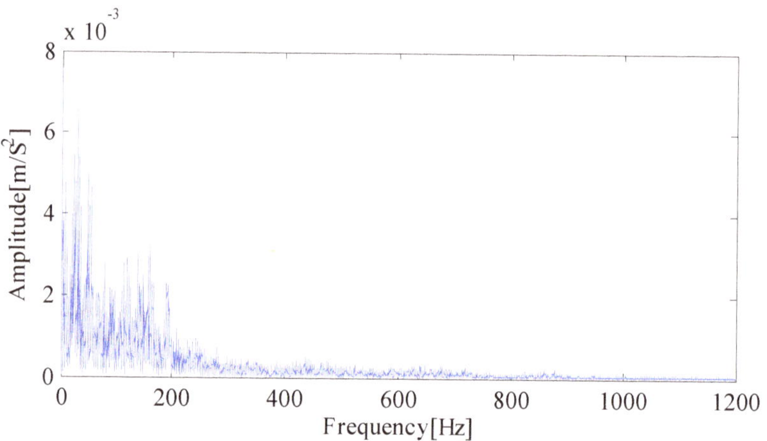

Figure 8. Frequency-domain waveform of check valve fault operation.

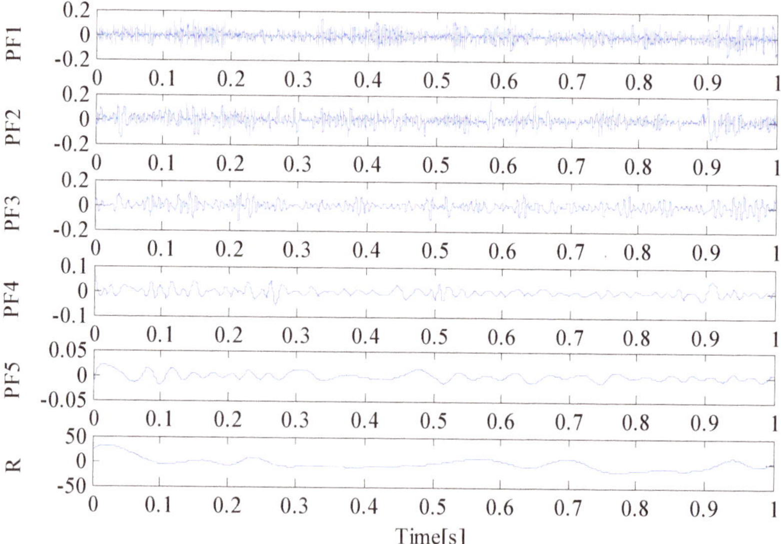

Figure 9. LMD decomposition results.

As shown in Table 1, the K-L divergence values of PF1, PF2, and PF3 were relatively small and less than the set threshold. Because the discrimination of K-L divergence was relatively obvious, it can be seen that their correlation with the original signal was relatively high. From the calculation results of kurtosis values, the kurtosis values of PF1 to PF5 were all greater than 3. Although it can be seen that the above component signals contained more impact components, this led to some difficulties in the signal screening. Therefore, the filtered PF1, PF2, and PF3 components were reconstructed according to the K-L divergence value. It can be seen from Table 2 that the K-L divergence values of IMF2, IMF 3, IMF4, and IMF 5 were less than the set threshold, which shows that their correlation with the original signal is relatively high. At the same time, by analyzing the calculation results of the kurtosis values from IMF 1 to IMF 10, it can be seen that the kurtosis values of other IMF components were greater than 3, except IMF 6, IMF 9, and IMF 10, which also made it difficult to screen effective signals. Similarly, the above component signal was selected for reconstruction according to the K-L divergence value calculation result. Although the above filtered signal components were highly correlated with the original signal, they contained a large number of fault signal features, and a large number of noise interference components. Therefore, to reduce the impact of noise on fault feature extraction, it was necessary to denoise the reconstructed signal by using the wavelet packet. In this experiment, the sym5 wavelet was used to decompose and reconstruct the reconstructed signal. The reconstructed signal waveform is shown in Figure 11, and the signal time-domain waveform after noise reduction is shown in Figures 12 and 13.

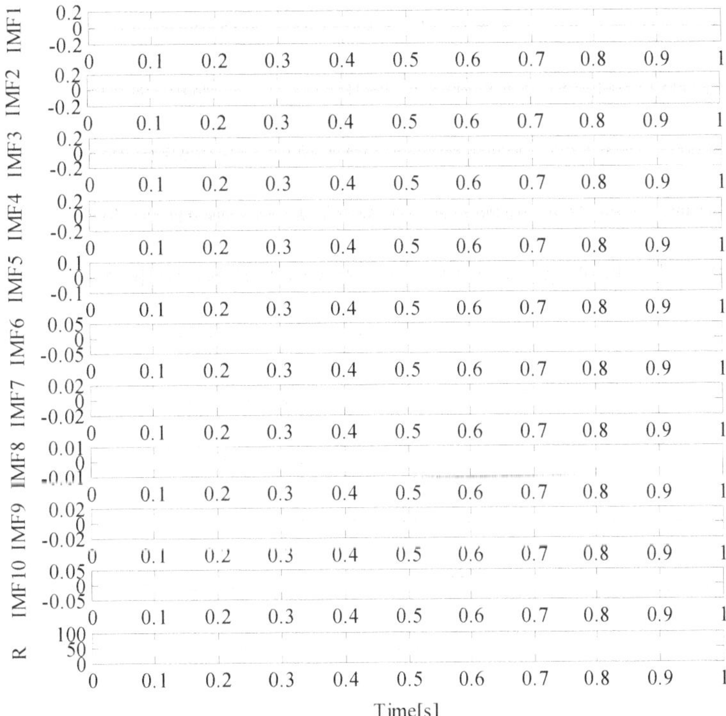

Figure 10. EMD decomposition results.

Table 1. K-L divergence and kurtosis of PF component.

Parameter	PF1	PF2	PF3	PF4	PF5
K-L divergence	0.009	0.001	0.026	0.253	1.001
Kurtosis	4.706	4.062	3.417	4.775	3.525

Table 2. K-L divergence and kurtosis of IMF component.

Parameter	IMF1	IMF2	IMF3	IMF4	IMF5
K-L divergence	0.036	0.007	0.009	0.001	0.016
Kurtosis	4.048	4.866	3.333	3.001	3.402
Parameter	**IMF6**	**IMF7**	**IMF8**	**IMF9**	**IMF10**
K-L divergence	0.111	0.561	1.001	0.724	0.234
Kurtosis	2.688	5.040	5.158	1.699	2.165

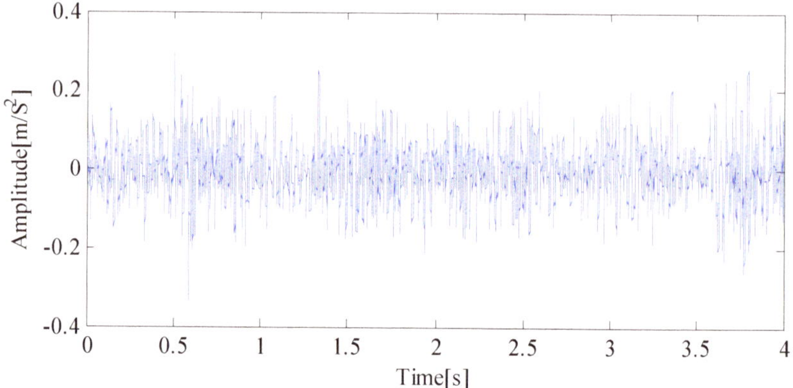

Figure 11. Time-domain waveform reconstruction.

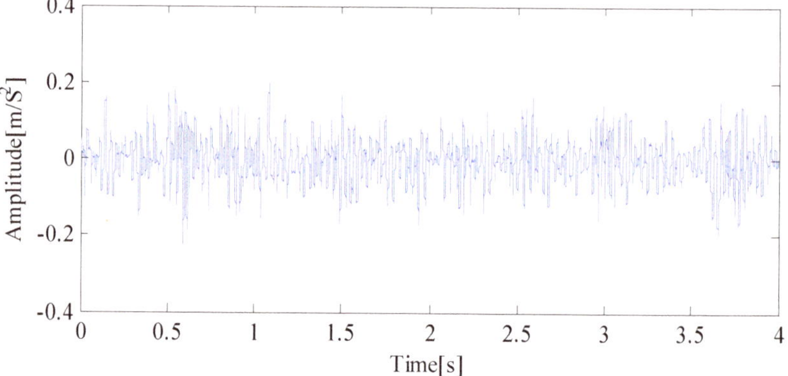

Figure 12. Time-domain waveform of signal based on LMD wavelet packet denoising.

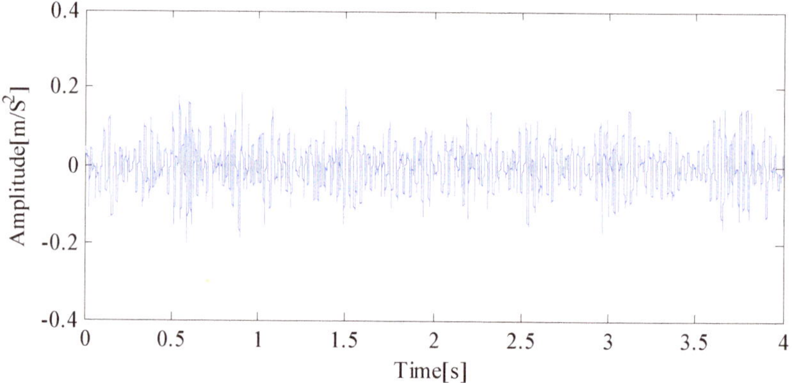

Figure 13. Time-domain waveform of signal based on EMD wavelet packet denoising.

Next, the signals based on LMD wavelet packet denoising, EMD wavelet packet denoising, and wavelet packet direct denoising were demodulated by the Hilbert envelope to compare and analyze the experimental results. The results are shown in Figures 14–16. It is evident from Figure 14 that there was a fundamental frequency (0.3125 Hz) and a second to sixth doubling frequency of the fundamental frequency components (0.625, 0.9375, 1.25,

1.563 and 1.875 Hz) in the envelope spectrum of the signal after noise reduction based on the LMD wavelet packet, which has become the dominant frequency of the vibration signal, indicating that a fault occurred at this time. As shown in Figure 15, fundamental frequency (0.3125 Hz) and other doubling frequency components appeared in the envelope spectrum of the signal denoised based on the EMD wavelet packet. Still, the overall amplitude was less than the result seen in Figure 14. As shown in Figure 16, frequency components such as fundamental frequency (0.3125 Hz) and the second doubling frequency (0.625 Hz) cannot be found in the envelope spectrum of the signal after wavelet packet noise reduction. Therefore, using the proposed method for fault feature extraction can achieve better results.

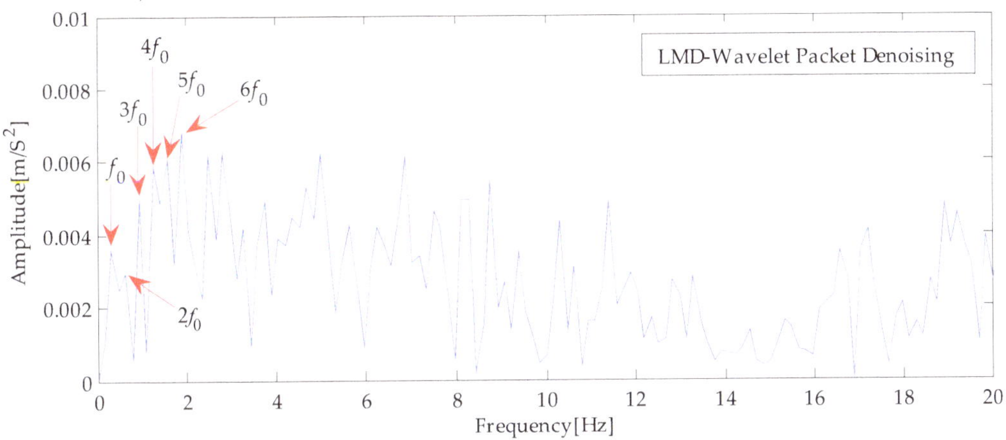

Figure 14. Signal envelope spectrum based on LMD wavelet packet denoising.

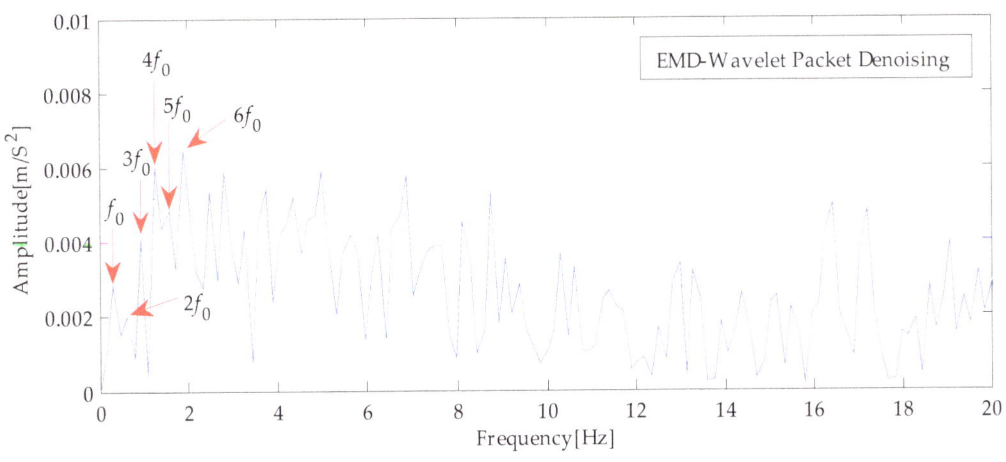

Figure 15. Signal envelope spectrum based on EMD wavelet packet denoising.

The above qualitative analysis shows that the proposed method has better fault feature extraction advantages than the other two methods. Firstly, the fault signal of the check valve was decomposed by the time-frequency analysis method, and the complete time-frequency distribution information of the signal was obtained. Compared with EMD, LMD suppresses the endpoint effect brought about by EMD, eliminates the problems of over envelope, under envelope, and mode aliasing caused by EMD, and the decomposed signal can retain the information of the original signal. Secondly, this research introduces K-L divergence as the screening criterion in the signal screening link. Through the calculation and comparative analysis of the K-L divergence value of each signal component, it can be seen that the discrimination of K-L divergence was relatively more significant, which is helpful to better select the signal component. Then, the subsequent reconstructed signal was denoised by a wavelet packet to filter out the noise interference. Finally, through the Hilbert envelope spectrum analysis, it can be seen that the characteristic frequency in the Hilbert envelope spectrum obtained by the proposed method was relatively apparent. Meanwhile, six frequency components such as the fundamental frequency of the fault and the second to sixth doubling frequency of the fundamental frequency can be extracted.

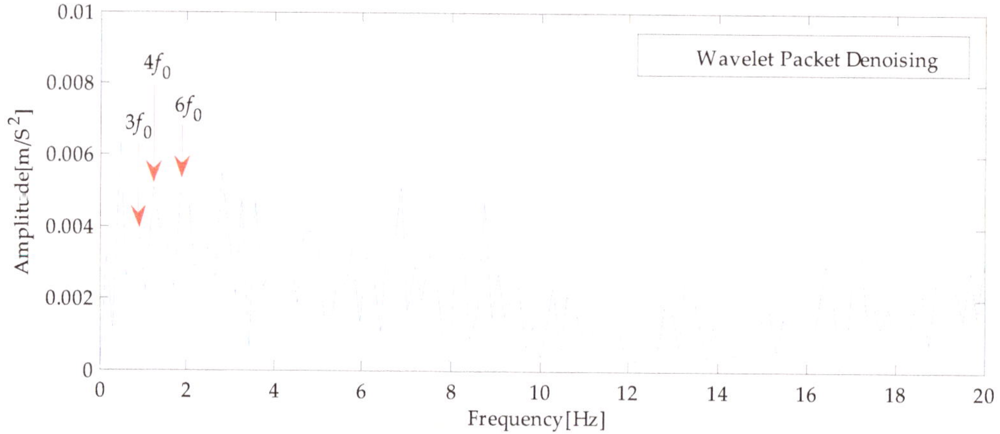

Figure 16. Signal envelope spectrum based on wavelet packet denoising.

7. Conclusions

Aiming at the non-stationary characteristics of the fault vibration signal of the check valve of a high-pressure diaphragm pump, a fault feature extraction method for the check valve based on LMD and wavelet packet analysis is proposed. Finally, the fault features were extracted by Hilbert envelope spectrum analysis. The following conclusions were obtained:

(1) The LMD method was used to decompose the original signal adaptively, which overcomes the problems of modal aliasing and endpoint effect caused by the EMD method, and there were fewer iterations. The component signal decomposed by this method only included five PF components and one residual signal component. In comparison, the EMD obtained more signal components (ten IMF components and one residual component). The results show that LMD can extract the time-frequency information of the signal more effectively and provide a guarantee for the screening of subsequent signal components.

(2) By calculating the K-L divergence value of each signal component and selecting the signal component whose K-L divergence value is less than the set threshold as the effective component signal, the problem of poor discrimination caused by the traditional kurtosis method can be avoided. Because the original signal contains more impact components, it is difficult to screen the signal. The experimental results in the signal screening link show that three signal components, PF1, PF2, and PF3, can be extracted as effective signals by using K-L divergence. If the kurtosis criterion is used for filtering, all signal component values are more significant than the set threshold and cannot be selected. At the same time, in the comparative experiment using EMD, four signal components were successfully selected as effective signals by using K-L divergence. When kurtosis was used as the screening criterion, it also showed the defect of poor discrimination.

(3) By further comparing and analyzing the envelope spectra after wavelet packet denoising based on different methods, it can be seen that the peak of characteristic frequency in the Hilbert envelope spectrum obtained by using the method proposed in this paper

is relatively apparent. The experimental results of envelope spectrum analysis show that six frequency components such as the fundamental frequency of check valve fault and the frequency from the second to sixth doubling frequency of the fundamental frequency can be extracted by using the method proposed in this paper. Although the above six frequency components were also extracted by the EMD wavelet packet joint denoising method, the overall amplitude was lower than that obtained by the method proposed in this paper. In addition, three frequency components such as 0.3125, 0.625 and 1.563 Hz cannot be found in the envelope spectrum of the signal denoised by wavelet packet alone. It is proved that the method proposed in this paper can extract the fault characteristics of the check valve more effectively.

As a large reciprocating industrial equipment, a high-pressure diaphragm pump can operate under complex working conditions such as high pressure, high temperature, high corrosion, and high concentration and is more widely used in the mining, metallurgy, petroleum, and chemical industries. This paper starts with the vibration signal analysis of the check valve, the core component of the high-pressure diaphragm pump, completes the fault feature extraction based on LMD and wavelet packet analysis, and provides a new idea for the research in this field. However, it may be affected by many uncertain factors and various fault forms under actual working conditions, so the proposed method may not be well applicable. Therefore, in future work, the research group will deeply summarize the latest theory and results of the time-frequency analysis method and further improve and perfect the proposed method to extract the operation state information of the check valve more effectively.

Author Contributions: Conceptualization, J.Y. and C.Z.; Data curation, J.Y.; Formal analysis, J.Y. and C.Z.; Funding acquisition, J.Y.; Methodology, J.Y.; Software, J.Y. and C.Z.; Validation, J.Y.; Writing—review and editing, C.Z. All authors have read and agreed to the published version of the manuscript.

Funding: This work was funded by the Scientific research fund project of Baoshan University (ZKMS202101), Special Basic Cooperative Research Programs of Yunnan Provincial Undergraduate Universities' Association (grant NO. 2019FH001-121), collaborative education project of industry university cooperation of the Ministry of Education (202102049026), 10th batches of Baoshan young and middle-aged leaders training project in academic and technical (202109), and the PhD research startup foundation of Yunnan Normal University (No. 01000205020503131).

Institutional Review Board Statement: Not applicable.

Informed Consent Statement: Informed consent was obtained from all subjects involved in the study.

Data Availability Statement: The data used to support the findings of this study are available from the corresponding author upon request.

Conflicts of Interest: The authors declare that they have no conflict of interest.

References

1. Zhang, H.; Chen, G.; Liu, L. Application of Slon high-gradient vertical-ring magnetic separator to remove iron from non-metal mineral. *Min. Eng.* **2004**, *2*, 43–46. (In Chinese)
2. Cattermole, K.W. The fourier transform and its applications. *Electron. Power* **2009**, *11*, 357–359. [CrossRef]
3. Griffin, D.; Lim, J.S. Signal estimation from modified short-time fourier transform. *IEEE Trans. Acoust. Speech Signal Process.* **1984**, *32*, 236–243. [CrossRef]
4. Georgakis, A.; Stergioulas, L.K.; Giakas, G. Wigner filtering with smooth roll-off boundary for differentiation of noisy non-stationary signals. *Signal Process.* **2002**, *82*, 1411–1415. [CrossRef]
5. Huang, N.E.; Shen, Z.; Long, S.R.; Wu, M.C.; Shih, H.H.; Zheng, Q.; Yen, N.-C.; Tung, C.C.; Liu, H.H. The empirical mode decomposition and the Hilbert spectrum for nonlinear and non-stationary time series analysis. *Proc. R. Soc. Lond. Ser. A Math. Phys. Eng. Sci.* **1998**, *454*, 903–995. [CrossRef]
6. Smith, J.S. The local mean decomposition and its application to EEG perception data. *J. R. Soc. Interface* **2005**, *2*, 443–454. [CrossRef]
7. Dong, L.; Jiang, R.; Xu, N.; Qian, B. Research on microseismic signal denoising method based on LMD–SVD. *Adv. Eng. Sci.* **2019**, *51*, 126–136. (In Chinese)

8. Wang, H.; Li, K.; Lian, J. Dynamic parametric identification for a hydropower house based on data fusion and LMD. *J. Vib. Shock* **2018**, *37*, 175–181.
9. Wang, Z.; Wu, W.; Ma, W.; Zhang, J.; Wang, J.; Li, W. Fault signal extraction method of rolling bearing weak fault based on LMD-MS. *J. Vib. Meas. Diagn.* **2018**, *38*, 1014–1020.
10. Gupta, P.; Singh, B. Investigation of tool chatter using local mean decomposition and artificial neural network during turning of Al 6061. *Soft Comput.* **2021**, *25*, 11151–11174. [CrossRef]
11. Liao, L.; Huang, B.; Tan, Q.; Huang, K.; Ma, M.; Zhang, K. Development of an improved LMD method for the low-frequency elements extraction from turbine noise background. *Energies* **2020**, *13*, 805. [CrossRef]
12. Huang, Y.; Wang, X.; Yang, D.; Wang, L.; Gu, J.; Zhang, X.; Wang, K. A Weld quality classification approach based on local mean decomposition and deep belief network. *J. Mater. Eng. Perform.* **2021**, *30*, 2229–2237. [CrossRef]
13. Lee, C.Y.; Zhuo, G.L. Effective rotor fault diagnosis model using multilayer signal analysis and hybrid genetic binary chicken swarm optimization. *Symmetry* **2021**, *13*, 487. [CrossRef]
14. Venkata, P.B.; Chinara, S. Automatic classification methods for detecting drowsiness using wavelet packet transform extracted time-domain features from single-channel EEG signal. *J. Neurosci. Methods* **2020**, *2020*, 108927.
15. He, K.; Xia, Z.; Si, Y.; Lu, Q.; Peng, Y. Noise reduction of welding crack AE signal based on EMD and wave-let packet. *Sensors* **2020**, *20*, 761. [CrossRef] [PubMed]
16. Wang, X. Optimized VMD-wavelet packet threshold denoising based on cross-correlation analysis. *Int. J. Perform. Eng.* **2018**, *14*, 2239–2247. [CrossRef]
17. Sun, W.; Xiong, B.S.; Huang, J.P.; Mo, Y. Fault diagnosis of a rolling bearing using Wavelet packet de-noising and LMD. *J. Vib. Shock* **2012**, *31*, 153–156.
18. Zhang, F.; Liu, Y.; Chen, C.; Li, Y.F.; Huang, H.Z. Fault diagnosis of rotating machinery based on kernel density estimation and Kullback-Leiblerdivergence. *J. Mech. Sci. Technol.* **2014**, *28*, 4441–4454. [CrossRef]
19. Hemmati, F.; Orfali, W.; Gadala, M.S. Roller bearing acoustic signature extraction by wavelet packet transform, applications in fault detection and size estimation. *Appl. Acoust.* **2016**, *104*, 101–118. [CrossRef]
20. Gómez, M.J.; Castejón, C.; Corral, E.; García-Prada, J.C. Railway axle condition monitoring technique based on wavelet packet transform features and support vector machines. *Sensors* **2020**, *20*, 3575. [CrossRef]
21. Shi, X.; Qin, P.; Zhu, J.; Xu, S.; Shi, W. Lower Limb Motion Recognition Method Based on Improved Wavelet Packet Transform and Unscented Kalman Neural Network. *Math. Probl. Eng.* **2020**, *2020*, 1–16. [CrossRef] [PubMed]
22. Zhou, C.; Jia, Y.; Bai, H.; Xing, L.; Yang, Y. Sliding dispersion entropy-based fault state detection for diaphragm pump parts. *Coatings* **2021**, *11*, 1536. [CrossRef]

Article

A Fault Diagnosis Method Based on EEMD and Statistical Distance Analysis

Tingzhong Wang [1,*], Tingting Zhu [1], Lingli Zhu [1] and Ping He [2]

[1] School of Information Technology, Luoyang Normal University, Luoyang 471934, China; muyixin2021@126.com (T.Z.); 13783141977@163.com (L.Z.)

[2] HIKVISION, Hangzhou 310052, China; 13083962903@163.com

* Correspondence: wangtingzhong@lynu.edu.cn

Abstract: Serious vibration or wear with large friction usually appear when faults occur, which leads to more serious faults such as the destruction of the oil film, bringing great damages to both the society and economic sector. Therefore, the accurate diagnosis of a fault in the early stage is important for the safety operation of machinery. To effectively extract the fault features for diagnosis, EMD-based methods are widely used. However, these methods spend lots of efforts diagnosing faults and require plenty of professional knowledge of diagnosis. Although many intelligent classifiers can be used to automatically diagnose faults such as wear, a broken tooth and imbalance, the combing EMD-based method, the scarcity of samplings with labels hinder the application of these methods to engineering. It is because the model of the intelligent classifier must be constructed based on sufficient samplings with a label. To solve this problem, we propose a novel fault diagnosis method, which is performed based on the EEMD and statistical distance analysis. In this method, the EEMD is used to decompose one original signal into several IMFs and then the probability density distribution of each IMF is calculated. To diagnose the fault of the machinery, the Euclidean distance between the signal acquired under an unknown fault with the other referenced signals acquired previously under various fault types is calculated. At last, the fault of the signal is the same with the referenced signal when the distance is the smallest. To verify the effectiveness of our proposed method, a dataset of bearings with different faults, and a dataset of 2009 Prognostics and Health Management (PHM) data challenge, including gear, bearing and shaft faults are used. The result shows that the proposed method can not only automatically diagnose faults effectively, but also fewer samplings with a label are used compared with the intelligent methods.

Keywords: fault diagnosis; rotating machinery; EEMD; probability density distribution; statistical distance

Citation: Wang, T.; Zhu, T.; Zhu, L.; He, P. A Fault Diagnosis Method Based on EEMD and Statistical Distance Analysis. *Coatings* **2021**, *11*, 1459. https://doi.org/10.3390/coatings11121459

Academic Editors: Ke Feng and Diego Martinez-Martinez

Received: 3 November 2021
Accepted: 26 November 2021
Published: 28 November 2021

Publisher's Note: MDPI stays neutral with regard to jurisdictional claims in published maps and institutional affiliations.

1. Introduction

Rotating machinery is widely used in industry and plays an important role in transmitting force, bearing load, etc. It usually operates under a tough environment and is subject to failure, which likely leads to significant economic losses or injuries. Therefore, it is necessary to monitor the health state of machinery and diagnosis defects timely and the research on fault diagnosis has attracted much attention in recent years. To inspect the condition of machinery, vibration sensors are always used for acquiring the vibration signal, which contains the health condition information. Unluckily, the machinery probably operates under a tough environment, so the useful information is always submerged in environment noise [1].

To extract the weak and useful information from the vibration signal under a low signal-to-noise ratio, many signal processing-based approaches have been proposed for fault diagnosis, such as the wavelet analysis [2], Kurtogram [3], Infogram [4] and singular value decomposition (SVD) [5]. Among these methods, the empirical mode decomposition (EMD) is one of the most used methods for fault diagnosis and has been used for gear fault diagnoses [6], bearing fault diagnoses [7] and abnormal clearance diagnoses of diesel engines [8], etc. The characteristics information of the fault can be effectively extracted using this method, because the EMD is able to decompose the complicated signals into several intrinsic model functions (IMFs), which represent the intrinsic oscillation modes embedded in the signal. Subsequently, the envelop spectrum technique can be applied to process the IMFs for finding the fault frequency. To overcome some shortcomings of the EMD, including negative frequency values, mode mixing and the lack of a rigorous mathematical formulation [9], many methods have been proposed. For example, an ensemble empirical mode decomposition (EEMD) [10], wavelet packet decomposition (WPD), local feature scale decomposition (LCD) [11], and BS-EMD [12] have also been proposed to solve the problem of mode mixing.

Although these EMD and EMD-based methods are good at extracting the impulsive signal generated by a machinery fault, it still spends lots of effort diagnosing the fault and require plenty of professional knowledge. To make full use of the merits of EMD and automatically classify the fault, many intelligent methods can be used as the classifier. Cheng et al. used the singular value decomposition (SVD) to initial feature vector matrices based on the EMD and then the support vectors machines (SVMs) trained by these features were used to distinguish the fault patterns of gears and roller bearings [13]. To achieve a better fault diagnosis performance, a more accurate SVM classifier named the multiclass transudative support vector machine (TSVM), trained by both time-domain features and time-frequency features extracted from IMFs, was applied to diagnose the gear faults [14]. Similarly, Zhang and Zhou proposed a novel procedure based on the EMD and optimized the SVM [15]. The extracted features include two types of features, the EEMD energy entropy and singular values of the matrix whose rows are IMFs. Yu et al. [16] improved the fault diagnosis performance of the SVM classifier by introducing the K-means method for selecting the most sensitive IMFs. The features, including nine time-domain features, were extracted from the most sensitive IMFs for training the SVM model.

Ali et al. [17] selected the most significant IMFs and then the chosen features, including the time features and time domain features, were used to train an artificial neural network (ANN) for classifying bearing defects. Bin et al. [18] combined the wavelet packet decomposition (WPD) and EMD to extract the fault features, and then these features were input into the classical three-layer BP neural network model for fault diagnosis. Xiao et al. [19] used an improved EMD energy entropy as the input of the SVM optimized by particle swarm optimization (PSO). A new neural network named the probability neural network (PNN) was also utilized to classify the fault models based on the features extracted from IMFs decomposed using the variational mode decomposition (VMD). The features consist of multifractal features extracted by MFDFA, and the generalized Hurst exponents and the dimension of these features were reduced by the principal component analysis (PCA) [20]. The neuro-fuzzy inference system (ANFIS) has been used as a classier for fault diagnosis by using a Fuzzy entropy as a feature [21]. Tian et al. [22] used the SVD to compress the product functions (PFs) decomposed using local mean decomposition (LMD) for obtaining the feature vectors and then these feature vectors were input into the extreme learning machine model. According to these references, it can be found that feature extraction from the decomposed components, such as IMFs or PFs, is a vital and necessary process for constructing the fault diagnosis model. Moreover, lots of samplings with labels are needed to train an accuracy model. However, it usually spends much effort extracting useful features from signals to train models and labeled signals in a practical application are scarce.

To solve this problem, a new fault diagnosis method is proposed based on the EEMD and statistic distance analysis. First, the probability density functions of the acquired signal to be analyzed are calculated. Specifically, in the proposed method, to overcome the shortcomings of the EMD, the EEMD instead of the EMD is used to decompose the vibration signal into several IMFs. The probability density function (PDF) of each IMF can then be calculated. Second, the Euclidean distance between the probability density functions (PDF) of the signal and the samplings with different fault labels can be calculated. The statistical distance (SD) can be evaluated by summing up all the distance. A small SD indicates the similarity of two signals (also small) and their corresponding machines are more likely to have the same fault type. Therefore, the fault can be diagnosed and the machines have the same fault type with the sampling when their similarity is the smallest. The SD is calculated using the probability density function of the IMFs directly and no time or time-frequency features are extracted, so it is more easy to realize the method. Moreover, because the referenced signal of the fault to be diagnosed should be provided, the number of the sampling with labels just equals the number of fault types which satisfies the requirement of a practical application. Eventually, the proposed method is more suitable for fault diagnosis, which is verified using two cases. The main contribution in this paper is that an EEMD-based method is proposed to diagnose faults more conveniently, which tries to perform fault diagnosis adaptively and with fewer training samplings than the intelligent fault diagnosis.

The rest of this paper is structured as follows: Section 2 introduces the proposed method in detail. Two cases are used to demonstrate the effectiveness of the proposed method in Section 3. Section 4 concludes the paper.

2. Proposed Diagnosis Model

In this section, the proposed diagnosis model was described. The proposed model consisted of four steps, including decomposing signals into IMFs using the EEMD, calculating the probability density distribution of different IMFs, computing the distance of the similarity between any two sampling, and determining the fault type using the similarity. The steps are described as follows in detail.

2.1. Decompose Signals into IMFs Using EEMD

First, signals of different types were decomposed into several IMFs using the EEMD, respectively. As a noise-assisted method, the EEMD could overcome the model mixing of the EMD.

For one signal $x(t)$ $(t = 1, 2, \ldots n)$, the EEMD could be calculated using the following six steps.

Step 1: Parameters used in the EEMD such as the trial number m and noise amplitude e were initialized

Step 2: The white noise x_m with the predefined amplitude was added to the signal $x(t)$ $(t = 1, 2, \ldots n)$ and the corresponding equation was given as:

$$x_m = x + n_m \tag{1}$$

where x_m and n_m represent the mth trial noise-assisted signal and added white noise, respectively.

Step 3: IMFs $imf_{i,m}$ could be obtained by decomposing x_m using the EMD algorithm for the mth trial noise-assisted signal, where i represents the ith IMFs.

Step 4: Steps 2–3 were repeated until $m = M$, but the white noise series were different between different repeats.

Step 5: The ensemble mean of the M trials was calculated for ith IMFs and shown as:

$$em_i = \frac{1}{M} \sum_{m=1}^{M} imf_{i,m} \tag{2}$$

Step 6: The ensemble mean could be calculated for each IMF and output the ensemble mean $em_i (i = 1, \ldots I)$ as the final IMF.

2.2. Calculate Probability Density Distribution of Different IMFs

Next, the probability density distribution of IMF em_i was calculated according to:

$$f_i(x) = \frac{1}{Nd} \sum_{j=1}^{N} K\left(\frac{x_j - x}{d}\right) \tag{3}$$

where d is the bandwidth and $d > 0$, N represents the whole number of serial points, x was taken from the corresponding IMF and $K(\cdot)$ denotes the non-negative kernel function which was selected as the Gaussian function in the proposed method. The Gaussian function was as follows:

$$g(x) = \frac{1}{\sqrt{2\pi}} e^{-\frac{(x_j - x)}{2d^2}} \tag{4}$$

2.3. Compute the SD for Evaluating Similarity between Any Two Samplings

Then, the similarity between any two IMFs (e.g., em_i and em_j) could be evaluated by computing the statistical distance of their probability density distributions using the following equation:

$$SD_{IMFs}(f_1, f_2) = \sum_{i=1}^{m} \left(\sum_{j=1}^{n} (f_1(x_{i,j}) - f_2(x_{i,j}))^2\right)^{1/2} \tag{5}$$

where $f_1(x_{i,j})$ represents the j th probability density value of ith IMF, m is the total number of IMFs and n is the total number of serial points in each IMF. Obviously, the similarity between two signals could be measured by this distance, which considered the difference of the probability density distribution at different IMFs. In particular, the more similar two signals were, the smaller the distance was.

2.4. Determine the Fault Type Using the Similarity

Signals generated by machinery with the same fault type has a large similarity. On the contrary, the similarity of signals generated by machinery with different fault types was much smaller than the same fault type. According to this rule, the fault could be diagnosed using the proposed statistical-based distance. Specifically, the SD between one signal with an unknown fault type and referenced signals with different fault types could be calculated. Then, the fault type of the signal could be determined using the following equation:

$$T = \underset{i}{argmin}\, D_{IMFs}(f_u, f_i) \tag{6}$$

where f_u is the probability density function of the signal with an unknown fault type and f_i represents the signal with the ith fault type. The signal had the same fault type with the referenced signal when their statistical distance was the smallest. It could be found that only one referenced signal was used, while lots of samples should be acquired to train an intelligent model. In other words, this proposed method needed much fewer samples than the intelligent fault diagnosis method. Furthermore, the proposed method was much easier to apply to the engineering cases, which had fewer parameters to tune than the intelligent fault diagnosis methods.

3. Experimental Demonstrations

Bears and gears are two critical components used to transfer force and moment in rotary machinery, and they are easily subject to failure. In this section, data collected from bearings and gears were used to verify the effectiveness of the proposed method, respectively.

3.1. Rolling Bearing Fault Diagnosis Based on the Proposed Method

The bearing dataset of the western reserve university [23] case was used to verify the effectiveness of our proposed method. Figure 1 shows the test rig for obtaining data of various bearing faults. A 1.49 Kw, three-phase induction motor, was used to provide power, and ball bearings were installed on the left of the motor to support the motor shaft. An accelerometer with a sampling of 12 kHz was installed on the house of the motor to collect vibration signals. In this demonstration, data of various faults under different degrees were used. Specifically, 12 health conditions included (1) normal ball fault with a fault severity. Single point faults were introduced to the test bearings using electro-discharge machining with fault diameters of 7 mils, 14 mils, 21 mils, 28 mils, and 40 mils (1 mil = 0.001 inches). The fault severity was evaluated by the defect size, including (2) 0.007 inches, (3) 0.014 inches, (4) 0.021 inches, and (5) 0.028 inches, an inner ring fault with a fault severity of (6) 0.007 inches, (7) 0.014 inches, (8) 0.021 inches, and (9) 0.028 inches, and an outer rolling fault with a fault severity of (10) 0.007 inches, (11) 0.014 inches, and (12) 0.021 inches. The vibration signals of each fault were divided into 20 segments and each segment had 5000 sampling points. These segments could be considered as different samples with different faults.

Figure 1. The bearing test rig.

These datasets with a total of 240 samples were used to verify the proposed method. In total, 12 samples with the whole fault types were used as the referenced data. The other samples were used to test the effectiveness of the proposed method. For example, the SD between the samples of the third condition (ball fault with a severity of 0.014) and the 12 referenced samples of different types were calculated. For instance, Figure 2 shows the statistical distance between the probability density distribution of 12 referenced samples between that of the whole normal samples. The SD of the same fault type was plotted using the same curve. The curve labeled with the blue cross plotted the SD between the data of a normal condition and the referenced data of a normal condition. Sample one was just the referenced sample, so its SD was zero. Furthermore, it can be found that the SD between the referenced sample with the whole samples had the smallest SD compared with the other referenced samples, which was plotted using the curve labeled with a blue cross. As a result, these samples could be classified as the type of the second condition, which was consistent with the fact. Similarly, the fault types of the other samples could be distinguished. The corresponding results can be seen from Figure 2j to Figure 2l. The accuracy of different fault types was calculated and presented in Figure 3. In Figure 2e, the number of the samples in testing was 20 and we calculated the SD of these samples with referenced signals of different faults. Obviously, the SD between each sample with the referenced signal of inner ring fault with a fault severity of 0.007 was the smallest. According to the proposed method, we could infer that the fault types of these samples were inner ring faults with a fault severity of 0.007. Obviously, the result was consistent with the fact, so the accuracy of the proposed method for Figure 2e was 100%. Of course, in some figures, the result was not very good, for example, Figure 2j, but the result was still higher than 60%. By calculating all the samples of different types, we could obtain the mean accuracy. It could be found that the mean accuracy of the whole samples was 87%, which was a satisfying result when the samples with labels were very scarce.

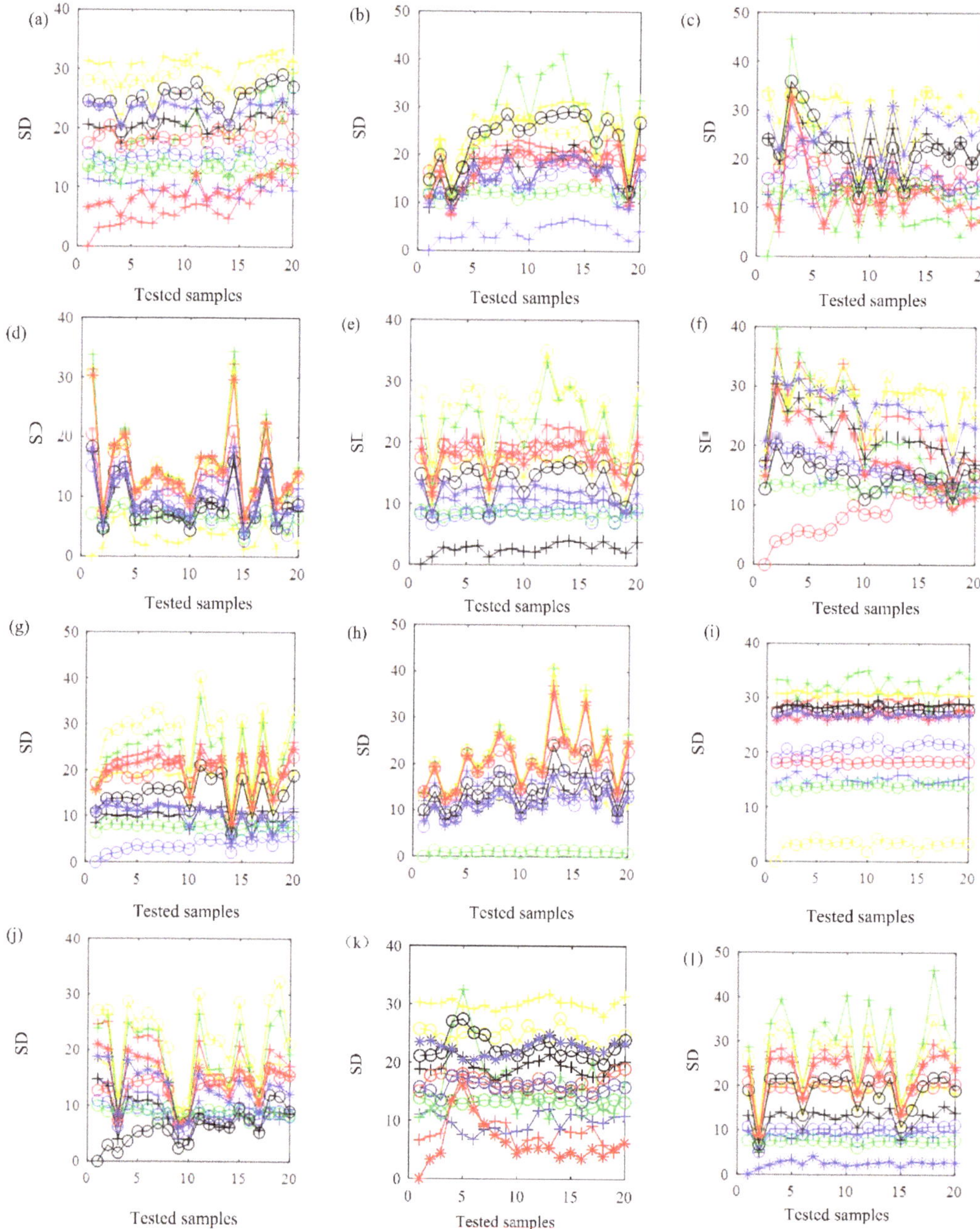

Figure 2. The SD between various referenced samples and different samples, including (**a**) ball fault with fault severity of 0.007, (**b**) ball fault with fault severity of 0.014, (**c**) ball fault with fault severity of 0.021, (**d**) ball fault with fault severity of 0.028, (**e**) inner ring fault with fault severity of 0.007, (**f**) inner ring fault with fault severity of 0.014, (**g**) inner ring fault with fault severity of 0.021, (**h**) inner ring fault with fault severity of 0.028, (**i**) normal, (**j**) outer rolling fault with fault severity of 0.007, (**k**) outer rolling fault with fault severity of (10) 0.014, and (**l**) outer rolling fault with fault severity of 0.021.

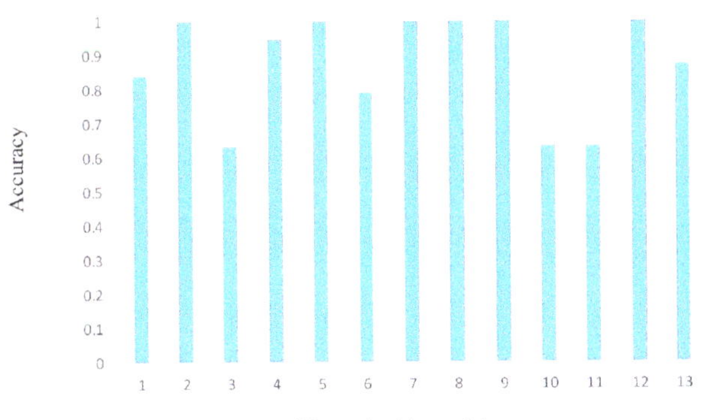

Figure 3. The diagnosis result of (1) ball fault with fault severity of 0.007, (2) ball fault with fault severity of 0.014, (3) ball fault with fault severity of 0.021, (4) ball fault with fault severity of 0.028, (5) inner ring fault with fault severity of 0.007, (6) inner ring fault with fault severity of 0.014, (7) inner ring fault with fault severity of 0.021, (8) inner ring fault with fault severity of 0.028, (9) normal, (10) outer rolling fault with fault severity of 0.007, (11) outer rolling fault with fault severity of 0.014, (12) outer rolling fault with fault severity of 0.021, and (13) the mean accuracy.

3.2. Fault Diagnosis of Gear Based on the Proposed Method

In this section, the vibration signals of spur gears with different faults were used to verify the effectiveness of the proposed method further. These data can be found in the 2009 PHM Challenge Competition [24]. Figure 4 shows the experiment setup which mainly consisted of one electric motor, four gears, six bearings, etc. The rotating frequency of the input shaft was 34.5 Hz and the tooth numbers of gears was 32, 96, 48, and 80, respectively. Two accelerometers were installed on both sides of the experiment setup to acquire the vibration data with a sampling frequency of 100/3 kHz. Different fault models such as NF (no fault), CH (chipped tooth), BR (broken tooth), ER (error), BS (ball spin fault), IR (inner race fault), OR (outer race fault), BA (bent axle), SI (shaft imbalance) and BK (bad key) were provided in this experiment. It was important to diagnose these faults as soon as possible, and if the equipment operated with faults, seriously vibrations would be generated and the friction of the surface between different components would become larger. The oil film would be destructed and more serious faults usually appeared.

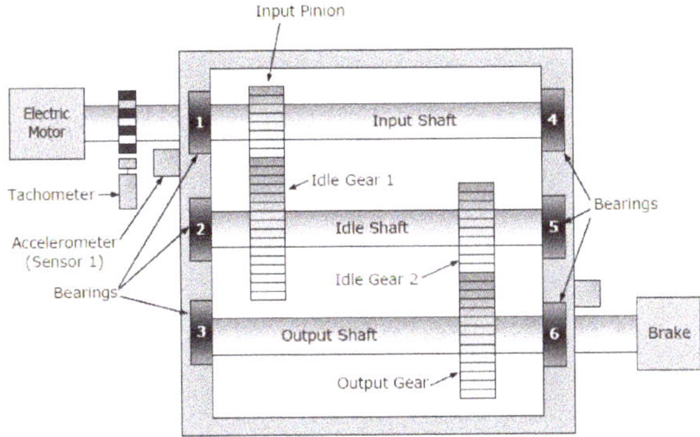

Figure 4. The experiment setup.

The final obtained dataset consisted of eight fault types. Concretely, these types were as follows: (1) normal, (2) input gear with CH and Idle 2 with ER, (3) Idle 2 with ER, (4) Idle 2 with ER, output with BR, and bearing one with BS, (5) input gear with CH, Idle 2 with ER, output with BR, bearing one with IR, bearing two with BS, bearing three with OR, (6) output with BR, bearing one with IR, bearing two with BS, bearing three with OR, Shafts input with SI, (7) bearing one with IR, shafts output with BK, and (8) bearing two with BS, bearing three with OR inputs(shafts) with BA.

Similar to case A, the different samples were obtained by dividing the vibration signals into different segments, each of which had 5000 segments. Therefore, a total of 320 samples could be obtained and eight samples with different fault types were considered as the referenced data in the proposed method. The statistical distance between the diagnosed samples with various faults and the referenced samples are plotted in Figure 5. The accuracy of all the diagnosis results was also calculated and the corresponding accuracy of various types as well as the mean accuracy can be seen in Figure 6. The mean accuracy of the whole samples was 87%, which also verified the effectiveness of the proposed method in fault diagnosing.

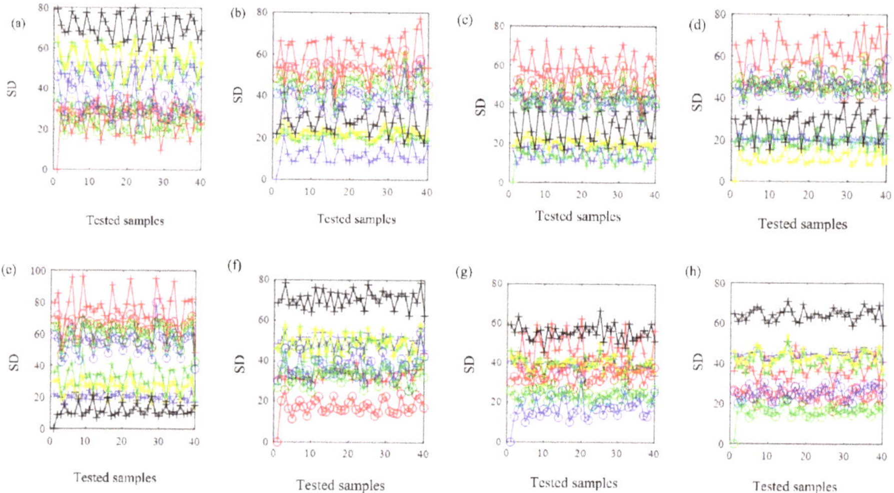

Figure 5. The SD between various referenced samples and different samples, including (**a**) normal, (**b**) input gear with CH and Idle 2 with ER, (**c**) Idle 2 with ER, (**d**) Idle 2 with ER, output with BR, and bearing 1 with BS, (**e**) input gear with CH, Idle 2 with ER, output with BR, bearing 1 with IR, bearing 2 with BS, bearing 3 with OR, (**f**) output with BR, bearing 1 with IR, bearing 2 with BS, bearing 3 with OR, Shafts input with SI, (**g**) bearing 1 with IR, shafts output with BK, and (**h**) bearing 2 with BS, bearing 3 with OR inputs(shafts) with BA.

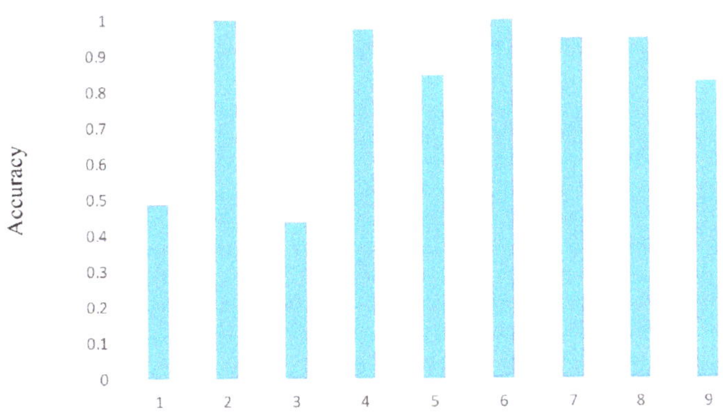

Figure 6. The diagnosis result of (1) normal, (2) input gear with CH and Idle 2 with ER, (3) Idle 2 with ER, (4) Idle 2 with ER, output with BR, and bearing 1 with BS, (5) input gear with CH, Idle 2 with ER, output with BR, bearing 1 with IR, bearing 2 with BS, bearing 3 with OR, (6) output with BR, bearing 1 with IR, bearing 2 with BS, bearing 3 with OR, Shafts input with SI, (7) bearing 1 with IR, shafts output with BK, (8) bearing 2 with BS, bearing 3 with OR inputs(shafts) with BA, and (9) the mean accuracy.

4. Conclusions

In this paper, an EMD-based statistic distance was proposed for the fault diagnosis of machinery. The proposed method made good use of the merit of the EEMD. Furthermore, a fault could be automatically diagnosed, which needed much fewer samples with labels than the intelligent method. The similarity of one signal between the referenced samples collected from the machinery with various faults could be accuracy evaluated. As a result, the signal had the same type with the referenced sample when they were most similar. The effectiveness of the proposed method was demonstrated by two real cases, including a bearing dataset and composite dataset of bearings and gears. It could be found that satisfying results were obtained even when there were a few samples with labels. Consequently, the proposed method can be further used in engineering, which will be considered in our further work.

Author Contributions: Data curation, T.Z.; Resources, P.H.; Software, P.H.; Writing—original draft, T.W.; Writing—review & editing, L.Z. All authors have read and agreed to the published version of the manuscript.

Funding: This work was supported in part by the Science and Technology Key Project of Henan province under Grant No. 212102210520 and 202102210370, and the Science and Technology Key Project in the High-tech Fields of Henan province, grant no. 152102210123.

Institutional Review Board Statement: Not applicable.

Informed Consent Statement: Not applicable.

Data Availability Statement: The data used to support the findings of this study are available from the corresponding author upon request.

Conflicts of Interest: The authors declare that they have no conflict of interest.

References

1. Lee, J.; Wu, F.; Zhao, W.; Ghaffari, M.; Liao, L.; Siegel, D. Prognostics and health management design for rotary machinery systems—Reviews, methodology and applications. *Mech. Syst. Signal Process.* **2014**, *42*, 314–334. [CrossRef]
2. Huo, Z.; Zhang, Y.; Francq, P.; Shu, L.; Huang, J. Incipient fault diagnosis of roller bearing using optimized wavelet transform based multispeed vibration signatures. *IEEE Access* **2017**, *5*, 19442–19456. [CrossRef]
3. Antoni, J. Fast computation of the kurtogram for the detection of transient faults. *Mech. Syst. Signal Process.* **2007**, *21*, 108–124. [CrossRef]
4. Antoni, J. The infogram: Entropic evidence of the signature of repetitive transients. *Mech. Syst. Signal Process.* **2016**, *74*, 73–94. [CrossRef]
5. Qiao, Z.; Pan, Z. SVD principle analysis and fault diagnosis for bearings based on the correlation coefficient. *Meas. Sci. Technol.* **2015**, *26*, 1–15. [CrossRef]
6. Li, Y.; Xu, M.; Wei, Y.; Huang, W. An improvement EMD method based on the optimized rational Hermite interpolation approach and its application to gear fault diagnosis. *Measurement* **2015**, *63*, 330–345. [CrossRef]
7. Tsao, W.-C.; Li, Y.-F.; Le, D.D.; Pan, M.-C. An insight concept to select appropriate IMFs for envelope analysis of bearing fault diagnosis. *Measurement* **2012**, *45*, 1489–1498. [CrossRef]
8. Li, Y.; Tse, P.; Yang, X.; Yang, J. EMD-based fault diagnosis for abnormal clearance between contacting components in a diesel engine. *Mech. Syst. Signal Process.* **2010**, *24*, 193–210. [CrossRef]
9. Feng, Z.; Zhang, D.; Zuo, M. Adaptive mode decomposition methods and their applications in signal analysis for machinery fault diagnosis: A review with examples. *IEEE Access* **2017**, *5*, 24301–24331. [CrossRef]
10. Jiang, H.; Li, C.; Li, H. An improved EEMD with multiwavelet packet for rotating machinery multi-fault diagnosis. *Mech. Syst. Signal Process.* **2013**, *36*, 225–239. [CrossRef]
11. Tong, Q.; Cao, J.; Han, B.; Zhang, X.; Nie, Z.; Wang, J.; Lin, Y.; Zhang, W. A fault diagnosis approach for rolling element bearings based on RSGWPT-LCD bilayer screening and extreme learning machine. *IEEE Access* **2017**, *5*, 5515–5530. [CrossRef]
12. Saidi, L.; Ali, J.B.; Fnaiech, F. Bi-spectrum based-EMD applied to the non-stationary vibration signals for bearing faults diagnosis. *ISA Trans.* **2014**, *53*, 1650–1660. [CrossRef] [PubMed]
13. Cheng, J.; Yu, D.; Tang, J.; Yang, Y. Application of SVM and SVD technique based on EMD to the fault diagnosis of the rotating machinery. *Shock Vib.* **2009**, *16*, 89–98. [CrossRef]
14. Shen, Z.; Chen, X.; Zhang, X.; He, Z. A novel intelligent gear fault diagnosis model based on EMD and multi-class TSVM. *Measurement* **2012**, *45*, 30–40. [CrossRef]
15. Zhang, X.; Zhou, J. Multi-fault diagnosis for rolling element bearings based on ensemble empirical mode decomposition and optimized support vector machines. *Mech. Syst. Signal Process.* **2013**, *41*, 127–140. [CrossRef]
16. Yu, X.; Dong, F.; Ding, E.; Wu, S.; Fan, C. Rolling bearing fault diagnosis using modified LFDA and EMD with sensitive feature selection. *IEEE Access* **2017**, *6*, 3715–3730. [CrossRef]
17. Ali, J.B.; Fnaiech, N.; Saidi, L.; Chebel-Morello, B.; Fnaiech, F. Application of empirical mode decomposition and artificial neural network for automatic bearing fault diagnosis based on vibration signals. *Appl. Acoust.* **2015**, *89*, 16–27.
18. Bin, G.F.; Gao, J.J.; Li, X.J.; Dhillon, B.S. Early fault diagnosis of rotating machinery based on wavelet packets—Empirical mode decomposition feature extraction and neural network. *Mech. Syst. Signal Process.* **2012**, *27*, 696–711. [CrossRef]
19. Xiao, Y.; Kang, N.; Hong, Y.; Zhang, G. Misalignment fault diagnosis of DFWT based on IEMD energy entropy and PSO-SVM. *Entropy* **2017**, *19*, 6. [CrossRef]
20. Liu, H.; Jing, J.; Ma, J. Fault diagnosis of electromechanical actuator based on VMD multifractal detrended fluctuation analysis and PNN. *Complexity* **2018**, *2018*, 1–11. [CrossRef]
21. Zheng, J.; Cheng, J.; Yang, Y. A rolling bearing fault diagnosis approach based on LCD and fuzzy entropy. *Mech. Mach. Theory* **2013**, *70*, 441–453. [CrossRef]
22. Tian, Y.; Ma, J.; Lu, C.; Wang, Z. Rolling bearing fault diagnosis under variable conditions using LMD-SVD and extreme learning machine. *Mech. Mach. Theory* **2015**, *90*, 175–186. [CrossRef]
23. Case Western Reserve University Bearing Data Center, "Bearing Data Center". Available online: http://csegroups.case.edu/bearingdatacenter/home (accessed on 15 July 2017).
24. PHM09 Data Challenge, "Prognostic Health Management Challenge". Available online: http://www.phmsociety.org/references/datasets (accessed on 28 September 2019).

Article

Sliding Dispersion Entropy-Based Fault State Detection for Diaphragm Pump Parts

Chengjiang Zhou [1,2], **Yunhua Jia** [1,2], **Haicheng Bai** [2,3,4,*], **Ling Xing** [1,2] and **Yang Yang** [1,2,3]

[1] School of Information Science and Technology, Yunnan Normal University, Kunming 650500, China; chengjiangzhou@foxmail.com (C.Z.); jia_yunhua@163.com (Y.J.); lin.xing@ynnu.edu.cn (L.X.); yangyang@ynnu.edu.cn (Y.Y.)
[2] The Laboratory of Pattern Recognition and Artificial Intelligence, Kunming 650500, China
[3] School of Physics and Electronic Information, Yunnan Normal University, Kunming 650500, China
[4] Network & Information Center, Yunnan Normal University, Kunming 650500, China
* Correspondence: 3701@ynnu.edu.cn

Abstract: Aiming at the disadvantages of low trend, poor characterization performance, and poor anti-noise performance of traditional degradation features such as dispersion entropy (DE), a fault detection method based on sliding dispersion entropy (SDE) is proposed. Firstly, a sliding window is added to the signal before extracting the DE feature, and the root mean square of the signal inside the sliding window is used to replace the signal in the window to realize down sampling, which enhances the trend of DE. Secondly, the hyperbolic tangent sigmoid function (TANSIG) is introduced to map the signals to different categories when extracting the DE feature, which is more in line with the signal distribution of mechanical parts and the monotonicity of the degradation feature is improved. For noisy signal, the introduction of locally weighted scatterplot smoothing (LOWESS) can remove the burrs and fluctuations of the SDE curve, and the anti-noise performance of SDE is improved. Finally, the SDE state warning line is constructed based on the 2σ criterion, which can determine the fault warning point in time and effectively. The state detection results of bearing and check valve show that the proposed SDE improves the trend, monotonicity, and robustness of the state tracking curve, and provides a new method for fault state detection of mechanical parts.

Keywords: mechanical parts; fault state detection; sliding dispersion entropy; feature extraction

1. Introduction

Diaphragm pump is a kind of transmission power equipment in the metallurgical industry, which provides power for slurry pipeline transmission. The safe operation of the diaphragm pump ensures the supply of mineral raw materials, and improves the production efficiency and the quality of steel products. The poor operating environment, stress, and load will cause damage to the diaphragm pump, which cause significant economic losses [1]. Therefore, the maintenance of the diaphragm pump is important. Bearing and check valve are the most frequently damaged parts in diaphragm pump, and the price of check valve is high. The maintenance personnel often detect the faults of bearing and check valve through abnormal sounds of diaphragm chamber and bearing seat, slurry leakage trace, pressure, and flow. These methods rely on subjective experience seriously. The excessive maintenance will cause the risk of shutdown, and the frequent replacement will cause the waste of spare parts, which will cause serious economic losses. Insufficient maintenance will lead to mechanical failure. Besides, a too late replacement of parts will lead to secondary failure of other parts, which will bring immeasurable losses and safety accidents. Therefore, it is urgent to propose a reliable fault detection method to guide the formulation of maintenance and replacement strategy.

Tracking the fault state of parts and determining the early fault point have important guiding significance for the design, assembly, and maintenance of the diaphragm pump.

Citation: Zhou, C.; Jia, Y.; Bai, H.; Xing, L.; Yang, Y. Sliding Dispersion Entropy-Based Fault State Detection for Diaphragm Pump Parts. *Coatings* **2021**, *11*, 1536. https://doi.org/10.3390/coatings11121536

Academic Editors: Ke Feng, Jinde Zheng and Qing Ni

Received: 1 November 2021
Accepted: 11 December 2021
Published: 14 December 2021

Publisher's Note: MDPI stays neutral with regard to jurisdictional claims in published maps and institutional affiliations.

The time domain, frequency domain, and time-frequency domain features of the vibration signal excited by faults will change in real time with the state degradation. However, it is difficult to track and detect the fault states of the diaphragm pump because the vibration signal is interfered by the pulsation of the slurry and the vibration of parts. In addition, the signal has nonlinear and non-stationary characteristics due to the influence of transmission path and hydraulic, mechanical, and electrical factors, which brings great challenges to the fault state tracking and detection.

There are few researches on fault detection of diaphragm pumps at home and abroad, but the researches on bearing fault detection still have good reference significance. Mixed domain features are the most commonly used methods in fault detection, including time domain, frequency domain, and time-frequency domain features. Li extracted 24 time-frequency features and selected sensitive feature through monotonicity and correlation coefficient, finally tracked the bearing degradation state through gate recurrent unit and 3σ criteria [2]. Gao extracted the degradation features from the mixed domain features of the bearing through isometric mapping, then established a reliability model by logistic regression [3]. Hua extracted the mixed domain features of the bearing and constructed a fault warning line based on the 3σ principle, and finally predicted the degradation state through the support vector machine [4]. Li selected the effective features from the mixed domain features of bearing and obtained a degradation curve by self-organizing feature mapping [5]. Bilendo extracted the mixed domain features and selected effective degradation features through local linear embedding (LLE) [6]. However, the single features, mixed domain features, and its fusion features have not achieved satisfactory results in the fault detection of the bearing and check valve. The reasons for this are that single features are only sensitive to the specific fault in a specific stage, the fusion features are redundant and depend on dimensionality reduction methods. Besides, the construction and dimensionality reduction of mixed domain features depend on the experience of technicians. The deep learning method can solve the above problems [7]. Ding proposed a domain adaptive long short-term memory (LSTM) to predict the bearing degradation state [8]. Hu extracted modes by convolutional neural network and evaluated the bearing degradation process by fuzzy C mean clustering [9].

Although the above-mentioned deep learning-based methods can avoid the subjective experience problems in mixed domains, the time cost of deep learning is higher and neural network structure seriously affects the feature extraction performance. Entropy can measure the complexity and uncertainty of signal [10] and has the advantages of simple calculation and high calculation efficiency [11]. Kumar extracted the Shannon entropy, permutation entropy (PE) and approximate entropy (AE) degradation features of the bearing and then constructed a bearing degradation trend model by gaussian process regression [12]. Noman first separated the oscillating eigenvalues from the vibration signal, then took the PE of the oscillation signal as the bearing degradation feature [13]. Minhas obtained several modes by empirical mode decomposition, then extracted the weighted multi-scale entropy degradation features of sensitive modes by Hurst index [14]. Li proposed a degradation feature combining composite spectrum and relative entropy, which can characterize the degradation trend of hydraulic pump effectively [15]. Mostafa tracked the fault evolution process of gear and bearing through sample entropy (SE), PE, and dispersion entropy (DE) [16]. Experiments show that the DE [17] is not sensitive to noise, but sensitive to instantaneous frequency, amplitude, and sequence bandwidth, which is in line with the feature extraction requirement of vibration signal.

However, the entropy methods also have their own defects, SE is sensitive to signal length, PE ignores the amplitude information, and they both have poor anti-noise performance. Although the performance of DE is slightly better than that of single features, mixed domain features and traditional entropy features in the fault detection of diaphragm pump, DE has not achieved satisfactory results in the fault detection. The problems can be summarized as follows. Firstly, the vibration signal segment used to extract the DE feature is discontinuous and irrelevant, which greatly reduces the tendency of the degradation feature. Secondly, the results obtained by the normal cumulative distribution function (NCDF) used in the DE deviate from the actual distribution of vibration signals, which makes the traditional DE feature unable to well characterize the true characteristics of vibration signals. Thirdly, the anti-noise performance of DE degradation feature still does not meet the fault detection requirements of mechanical equipment in the actual industrial environment. In addition, there is still a lack of an effective fault point detection and early warning method, which can track the degradation state of parts in real time and warn the key fault points.

The sliding dispersion entropy (SDE) and its state warning line are proposed and used for fault state detection and degradation state tracking in this paper. In order to enhance the tendency of DE degradation feature, a sliding window is added to the signal segment and the root mean square of the signal in the window is used to replace the signal segment to achieve down-sampling. In order to improve the characterization performance of DE feature, the down-sampling sequences are mapped to different categories by introducing hyperbolic tangent sigmoid function (TANSIG) mapping. Because the TANSIG mapping is closer to the actual empirical distribution of vibration signal, the proposed SDE enhances the monotonicity of the degradation feature. To enhance the anti-noise performance, locally weighted scatterplot smoothing (LOWESS) is introduced to remove the small fluctuations and burrs of the SDE feature curve. At the same time, an adaptive early warning line based on 2σ criteria is proposed, which can determine the fault warning point effectively. In summary, the proposed method solves the above problems well. The method can track the fault state of the parts and determine the fault warning point and provide technical guidance for the maintenance and replacement of the parts.

The remainder of this paper is organized as follows: in Section 2, the theory of the proposed sliding dispersion entropy (SDE) and its fault state warning line are introduced. Then, a state detection method based on SDE is proposed, and the specific steps are described in detail. In Section 3, the effectiveness of the SDE and its state warning line is proved by analyzing the bearing data in the laboratory environment, then the proposed SDE is applied to the fault detection of check valve in the actual industrial environment, and the proposed SDE was compared with many existing methods. Finally, some conclusions are presented in Section 4.

2. Methodology

2.1. Sliding Dispersion Entropy (SDE)

The degradation process of mechanical parts lasts a long time, and the categories and boundaries of fault states are fuzzy. Therefore, it is difficult for traditional features to characterize the degradation trend of parts [18]. DE can measure the complexity and chaotic characteristics of the signal, but its performance is not very good in tracking the state of check valve and bearing. The trend of vibration data is not fully considered in DE feature, and the description of vibration data distribution characteristics by DE is not accurate enough due to the use of normal cumulative distribution function (NCDF) mapping. In addition, DE is easily disturbed by small fluctuations and noise, and the reliability of tracking is poor. Therefore, a sliding dispersion entropy (SDE) based on sliding window down-sampling and TANSIG mapping is proposed. Assuming that the vibration signal of the mechanical part at the k-th ($k = 1, 2, \cdots, K$) time point has been collected, the SDE feature of the vibration signal at the current time point can be expressed as SDE_k, and its calculation procedures are as follows.

(1) In order to enhance the trend of the vibration signal obtained at the k-th time point, a sliding window with length P is added to the signal to be analyzed at first, and then the original signal segment in the sliding window is replaced by the root mean square (RMS) value of the segment signal. In this way, down-sampling is achieved. In the sliding process of the sliding window, the sliding window reaches the next window after sliding

$h = (P/2)$ sampling points each time. Wherein, the RMS of the signal segment in the i-th sliding window can be obtained by the following formula.

$$v_{rms}(i) = \sqrt{\sum_{m=(i-1)h+1}^{(i-1)h+P} u^2(m)/P},$$ (1)

$$v_{pre}(i) = v_{rms}(i) - \left(\sum_{i=1}^{N} v_{rms}(i)/N \right),$$ (2)

where $N = round(M/P)$ represents the total number of windows, M represents the number of data points at the k-th time point, and $v_{pre}(i)$ is the removing mean processing of the RMS values of the signal in the i-th window. Figure 1 shows the sliding window of the check valve signal when the signal length in the sliding window is $P = 2000$. In this paper, $P = 20$. The down-sampled signal $v_{pre}(i)(i = 1, 2, \cdots, N)$ of the signal in the N sliding windows can be obtained according to the above formula.

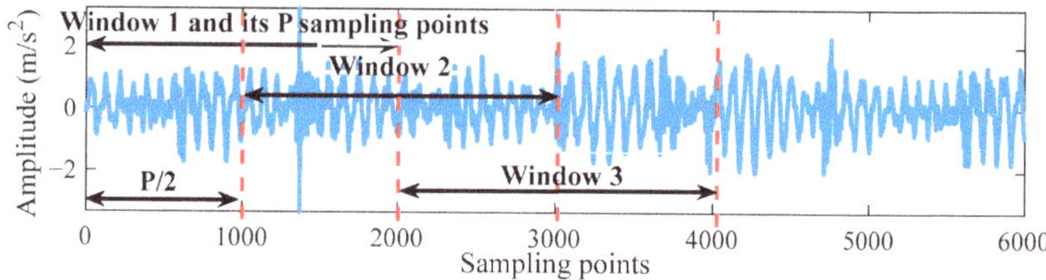

Figure 1. The schematic diagram of sliding window and down−sampling.

(2) In order to improve the characterization performance of DE, the down-sampling signal are mapped to different categories by introducing TANSIG mapping, and it is used as the SDE_k feature of the k-th time point signal. Its calculation is summarized as follows.

Step 1: Map the signal $x_j(j = 1, 2, \cdots, N)$ to c categories. Different from DE, the NCDF function of DE is replaced by the tan-sigmoid mapping function (TANSIG), after which the original signal x_j is mapped to the range between 0 and 1, i.e., $y = \{y_1, y_2, \cdots, y_N\}, y \in (0, 1)$. When the maximum or minimum value of sequence x deviates far from its mean or median, most of the data in sequence x is easy to be classified into a few categories, but TANSIG can solve the above problems well.

$$y_j = \frac{2}{1 + e^{-2\frac{(x_j - \mu)}{\sigma}}} - 1,$$ (3)

where σ and μ are the standard deviation and mean of sequence x. In order to verify the effectiveness of TANSIG function, we have compared and analyzed the mapping effect and distribution of linear function (LM), log sigmoid function (LOGSIG), TANSIG function, and NCDF function based on 76,800 data points of check valve fault signal. At the same time, these functions are also compared with the actual empirical distribution function (EDF) of the check valve data. As shown in Figure 2a, the curve corresponding to the TANSIG mapping function almost coincides with the curve corresponding to the EDF function and the check valve data distribution curve, which shows that the distribution of most check valve data is closer to the TANSIG mapping function.

Then, the vibration data points are transformed into the range of 1 to c by the linear transformation $c \cdot y_j + 0.5$, and then each y_j is classified into classes 1 to c according to the operation rule $z_j^c = round(c \cdot y_j + 0.5)$, where z_j^c refers to the j-th sequence point that has been classified and the symbol $round(\cdot)$ refers to rounding operation. Assuming the number of categories is 3 ($c = 3$), the classification of the first 2000 data points of the check valve fault signal is shown in Figure 2b, which indicates that the distribution of the vibration signal is closer to the actual empirical distribution of the data after being mapped by TANSIG.

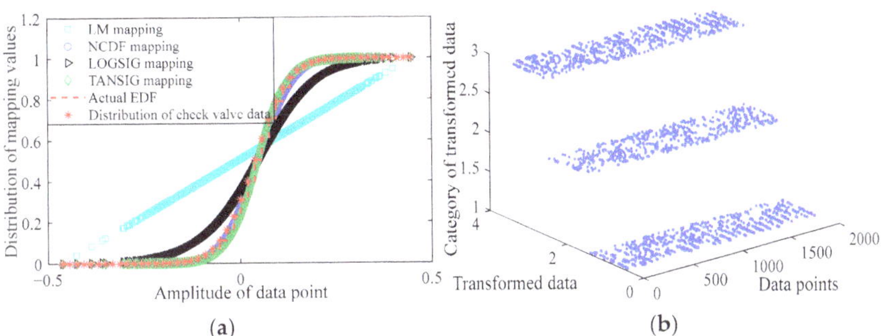

Figure 2. Comparison of mapping methods and classification of mapped data. (a) Comparison of different mapping methods; (b) data classification after TANSIG mapping.

Step 2: According to $z_i^{m,c} = \left\{ z_i^c, z_{i+d}^c, \cdots, z_{i+(m-1)d}^c \right\}$, each embedding vector $z_i^{m,c}, i = 1, 2, \cdots, N - (m-1)d$ is established by embedding dimension m and time delay d. Each $z_i^{m,c}$ is mapped to a dispersion pattern $\pi_{v0\cdots vm-1}$, where $z_i^c = v_0, z_{i+d}^c = v_1, \cdots, z_{i+(m-1)d}^c = v_{m-1}$, so the number of potential dispersion patterns that can be mapped to each $z_i^{m,c}$ is c^m. It can be seen that the sequence $z_i^{m,c}$ has m elements and each of them comes from an integer from 1 to c [19].

Step 3: For each potential dispersion mode $\pi_{v0\cdots vm-1}$ among the c^m dispersion modes, the relative dispersion frequency is $p(\pi_{v0\cdots vm-1})$, where $N - (m-1)d$ is the total number of embedded vectors.

$$p(\pi_{v0\cdots vm-1}) = \frac{Number\left\{ i \mid i \leq N - (m-1)d, z_i^{m,c} \quad has \quad type \quad \pi_{v0\cdots vm-1} \right\}}{N - (m-1)d}. \tag{4}$$

Step 4: Calculate DE according to the definition of Shannon entropy [10].

$$DE(x, m, c, d) = -\sum_{\pi=1}^{c^m} p(\pi_{v0\cdots vm-1}) \cdot \ln(p(\pi_{v0\cdots vm-1})). \tag{5}$$

The standardized dispersion entropy can be defined as $NDE(x, m, c, d) = DE(x, m, c, d)$ / $\ln(c^m)$, and relevant theories of DE can be found in literature [17].

(3) With the increase of the time point of data acquisition, that is, from the first time point to the k-th time point, and then from the k-th time point to the k-th time point, the SDE feature curve that characterize the state evolution process can be obtained by repeating the above step (1) and step (2). In the case of weak noise, the SDE feature curves can track the evolution process of mechanical operation state in real time.

(4) Under the condition of strong noise in the slurry transportation environment, there are burrs and random fluctuations in the SDE feature curve. Therefore, we can remove small fluctuations in the SDE feature curve by introducing locally weighted scatterplot smoothing (LOWESS). Each smoothing point of the SDE curve can be determined by adjacent data points within a given range, where the regression weight of the data points in a given range can be represented as follows.

$$\omega_i = \left(1 - |(x - x_i)/d(x)|^3 \right)^3, \tag{6}$$

where x represents the coordinates of the points that need to be smoothed, x_i represents the nearest neighbor point of x in a given range, and $d(x)$ is the distance between x and the furthest predicted value in a given range. Then, the initial weight is used to estimate and the robust coefficient is defined by the residual r_i.

$$\delta_i = \left(1 - \left|\frac{r_i}{6 Median(|r_1|, |r_2|, \cdots, |r_n|)}\right|\right). \tag{7}$$

The weight function is modified by iterating N times through the above steps, and the smooth value can be obtained according to the polynomial and weight. More theories about LOWESS can be found in reference [20]. The SDE method based on LOWESS is called smooth SDE in this paper.

In order to verify the effectiveness of SDE features, a mixed evaluation index (MEI) based on monotonicity, robustness and trend indexes is constructed and used to evaluate the state tracking performance of degraded features such as SDE.

Firstly, the mechanical parts are damaged gradually during the service period, and the degradation process of the fault state is irreversible except for repairs, so the degradation feature should be monotonous.

$$Mon(X) = \frac{1}{K - 1} \left|No. \quad of \quad d/dx > 0 - No. \quad of \quad d/dx < 0\right|, \tag{8}$$

where $X = \{x_k\}_{k=1:K}$ represents the feature sequence, x_k is the feature value corresponding to the time point t_k, and $d/dx = x_{k+1} - x_k$ represents the sequence gradient. The greater the $Mon \in [0, 1]$, the better the monotonicity of the feature curve, but Mon may fail when the curve fluctuates greatly.

Secondly, the vibration signal affected by environmental noise and operating conditions is non-stationary, and the random fluctuation of the degradation feature will reduce the reliability of state tracking, so the robustness index is used to evaluate the robustness of the feature.

$$Rob(X) = \frac{1}{K} \sum_{k=1}^{K} \exp\left(-\left|\frac{x_k - x_k^T}{x_k}\right|\right), \tag{9}$$

where x_k^T is the average trend value of the degradation feature at the time point t_k, which can be obtained through LOWESS. Finally, with the increase of degradation time, the degradation trend of mechanical parts becomes more and more obvious, so the trend index is used to evaluate the correlation between degradation feature and time.

$$Tre(X, T) = \frac{K\left(\sum_{k=1}^{K} x_k t_k\right) - \left(\sum_{k=1}^{K} x_k\right)\left(\sum_{k=1}^{K} t_k\right)}{\sqrt{\left[K\sum_{k=1}^{K} x_k^2 - \left(\sum_{k=1}^{K} x_k\right)^2\right]\left[K\sum_{k=1}^{K} t_k^2 - \left(\sum_{k=1}^{K} t_k\right)^2\right]}}. \tag{10}$$

The smaller the difference between the absolute value of $Tre(X, T) \in [-1, 1]$ and 1, the stronger the correlation between the degradation feature and time. In practice, it is difficult for a single index to evaluate the tracking ability of a degradation feature comprehensively, so a mixed evaluation index (MEI) is constructed in this paper.

$$MEI(X) = \alpha_1 Mon(X) + \alpha_2 Tre(X) + \alpha_3 Rob(X), \quad s.t. \quad \alpha_i > 0, \quad \sum_{i=1}^{3} \alpha_i = 1, \tag{11}$$

where α_1, α_2 and α_3 are 0.5, 0.3, and 0.2, respectively.

2.2. State Warning Line Based on SDE

Assuming that there is a set of random variables that approximately accord with normal distribution, and the mean of the set of random variables is μ and the variance is σ^2, then the probability that the variable is distributed in the range $(\mu - 3\sigma, \mu + 3\sigma)$ is 99.74%, and the probability that the variable is distributed in the range $(\mu - 2\sigma, \mu + 2\sigma)$

is 95.44%. If the variable is out of the range $(\mu - 2\sigma, \mu + 2\sigma)$, it is considered that the state has changed. Therefore, this criterion is introduced into fault detection and used to construct the adaptive threshold of the fault state warning line. If several consecutive SDE feature values exceed the range $(\mu - 2\sigma, \mu + 2\sigma)$, it is considered that the operating state of the mechanical parts has changed significantly, and the fault warning signal should be sent at this time. According to the SDE feature value of the vibration signal at the k-th $(k = 1, \cdots, K)$ time point, a time-varying adaptive state warning line can be constructed, and the adaptive warning line is shown below.

$$T_h(t) = mean(SDE(1 : t_k)) - 2std(SDE(1 : t_k)), \quad k = 1, 2, \cdots, K. \tag{12}$$

Assuming that the time period $(1 : t_k), k = 1, \cdots, s_1$ is the normal stage and the state does not change in this stage, we first calculate the lower threshold of degraded state curve $SDE_k, k = 1, \cdots s_1$. If it is necessary to determine whether the fault state changes at time point s_2, we only need to compare the $SDE(1 : t_{s_2})$ at time point s_2 with the threshold $T_h(t_{s_1})$ obtained at time point s_1, where time point s_2 is the next time point after time point s_1. If the $SDE(1 : t_{s_2})$ is within the threshold range $T_h(t_{s_1})$, it indicates that the fault state has not changed much. If the SDE values of consecutive multiple time points exceed the corresponding adaptive threshold, it indicates that the fault state has changed, and the first point exceeding the threshold is determined as the fault early warning point. In this way, an effective adaptive fault state early warning line can be obtained.

2.3. Fault State Detection Method

In the degradation process from normal to failure, the operating state of mechanical parts such as bearings and check valves has gone through several degradation stages. In order to detect the fault stages and detect fault state warning point accurately, a fault state detection and evaluation method based on SDE is proposed. The implementation steps for this method are shown in Figure 3.

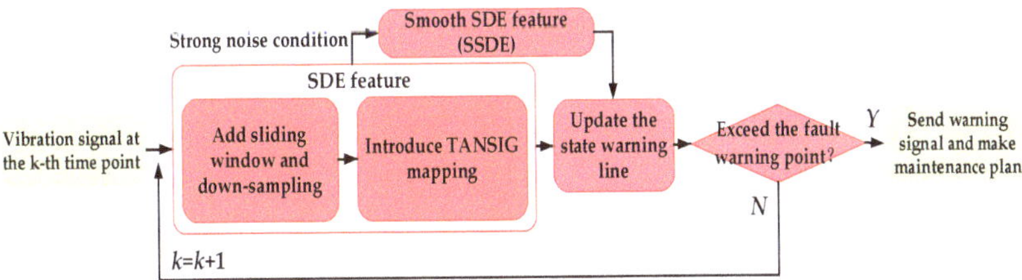

Figure 3. Flow chart of fault state detection and evaluation method.

(1) The vibration signal of the mechanical parts at the k-th time point is collected by interval sampling, and finally the vibration signal of the whole life cycle from normal to failure is obtained.

(2) Extract the first N data points from the vibration signal at the k-th time point and extract the SDE features of these data points. If there are too many burrs in the SDE feature curve, the smooth SDE feature of these data points are extracted.

(3) Update the state warning line in real time to detect whether the current operating state of mechanical parts has changed. If the fault warning point is not detected by the continuously updated state warning line, that is, there is no intersection between the state warning line and the SDE feature curve, then we continue to extract the SDE feature of the vibration signal at the next time point and update the state warning line according to Formula (12). The above process will continue until the state warning point is detected.

(4) If the state warning line crosses the fault warning point, that is, the value of the warning line at consecutive multiple time points crosses the SDE feature curve, the fault warning signal should be issued immediately and the corresponding maintenance plan should be formulated immediately.

In addition, in order to verify the effectiveness of the proposed SDE and adaptive state warning line, we constructed a mixed evaluation index (MEI) to evaluate the performance of fault state tracking, and compared SDE with single features, fusion features and traditional entropy features.

3. Results and Discussion

3.1. Bearing Fault Detection and Comparative Analysis

In order to verify the effectiveness of the proposed SDE feature and state warning line in the fault state detection of mechanical parts, we first analyzed the degenerate state of the bearing in the laboratory environment and compared the proposed method with single features, fusion features, and traditional entropy features. The bearing data from the intelligent maintenance system (IMS) of the University of Cincinnati is used as the experimental data in this paper, and the experimental platform is shown in Figure 4. Four sets of double-row Rexnord ZA-2115 roller (REXNORD, Milwaukee, WI, USA) bearings operate continuously at a speed of 2000 r/min under the action of the spring radial load of 26.671 kN. Among them, the diameter of the roller element $D_m = 71.501$ mm, the pitch diameter $d_r = 0.407$ mm, the number of roller elements $n_r = 16$, the contact angle $\theta = 15.17°$ PCB353B33 acceleration sensors (PCB PIEZOTRONICS, Buffalo, NY, USA) are fitted to each bearing shaft, and the vibration signal is collected once every 10 min via DAQ-6062E data acquisition card (National Instruments, Austin, TX, USA). The sampling frequency is 20 kHz and the data length is 20,480 points.

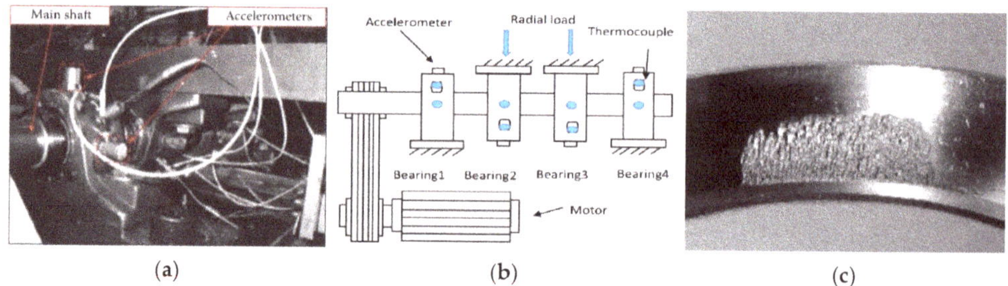

Figure 4. Illustration of the bearing experiment platform. (**a**) Bearing test rig; (**b**) sensor placement illustration; (**c**) outer race defect of set number 2.

In the data acquisition experiment, 984 sets of outer race vibration signals from 10:32:39 on 12 February to 06:22:39 on 19 February were collected, which means that the outer ring vibration signals lasting 164 h from normal to fault can be used to verify the effectiveness of the method. The experimental signal is from Bearing1 of 2nd_test, and the outer race defect as shown in Figure 4c. For the vibration signal corresponding to 984 time points, the first 2048 data points of the vibration signal at each time point are extracted respectively, and the obtained sampling signal is shown in Figure 5. The amplitude of the sampled signal increases gradually, but it is difficult to track and detect the fault state of the bearing based on the signal amplitude alone.

As the fault state of mechanical parts changes, the time domain amplitude and probability distribution, frequency domain energy and spectral peak position, time-frequency domain characteristics, and energy will change accordingly. Therefore, the traditional time domain, frequency domain and time-frequency domain features are often used to track the fault state of mechanical parts and its evolution process. In addition, a single feature can only characterize the characteristics of a specific fault in a specific fault stage. Therefore, feature fusion methods are gradually used for state tracking of mechanical parts, including principal component analysis (PCA), local linear embedding (LLE), linear local tangent space alignment (LLTSA), and so on.

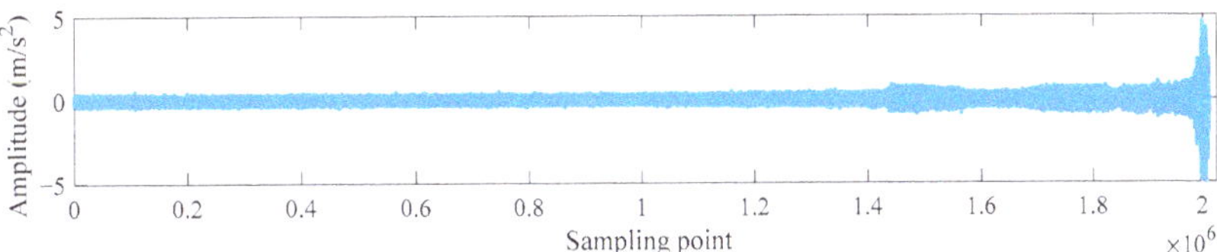

Figure 5. Bearing sampling signal.

In order to verify the state tracking performance of the proposed SDE feature, it is compared with time domain, frequency domain, time-frequency domain features, fusion features and entropy features. In the experiment, the values of $\alpha_1, \alpha_2, \alpha_3$ in the mixed evaluation index (MEI) are 0.5, 0.3, and 0.2, respectively [21]. For the vibration signals from the first time point to the 984th time point, the SDE features of the vibration signals at each time point are extracted respectively, and an SDE degradation curve is obtained finally. Similarly, for the vibration signals corresponding to 984 time points, their 16 time domain features (T1~T16), 13 frequency domain features (F1~F13), 8 wavelet packet energy features (TF1~TF8, 3-layer db5 wavelet), 3 mixed domain fusion features based on PCA, LLE and LLTSA, and 3 entropy features based on sample entropy (SE), permutation entropy (PE), and dispersion entropy (DE) can be obtained, and the state tracking performance of all the above degradation feature curves can be evaluated by the mixed evaluation index (MEI). As shown in Table 1, the MEI 0.4717 of the SDE feature is the largest, and the trend index (Tre) 0.8166 and robustness index (Rob) 0.9958 of the SDE feature are close to the corresponding maximum 0.8166 and 0.9999, which indicates that the SDE feature has the best state tracking performance. The MEI indices of F10 feature and LLE fusion feature are close to that of SDE feature, and none of the remaining degradation features can fully consider the monotonicity, robustness and trend of the degradation curve in the state tracking process.

Table 1. The evaluation results of bearing fault state detection.

Feature	Mon	Rob	Tre	MEI	Feature	Mon	Rob	Tre	MEI
T1	0.0193	0.9628	0.6334	0.4252	F7	0.0030	0.9857	0.7152	0.4403
T2	0.0010	0.9640	0.6269	0.4151	F8	0.0050	0.9929	0.6680	0.4340
T3 RMS	0.0233	0.9594	0.6301	0.4255	F9	0.0030	0.9898	0.7144	0.4413
T4	0.0193	0.9628	0.6334	0.4252	F10	0.0417	0.9720	0.7189	0.4562
T5	0.0010	0.8725	0.1857	0.2994	F11	0.0193	0.9696	0.0374	0.3080
T6 Kurt	0.0091	0.8050	0.1129	0.2686	F12	0.0050	0.9689	0.0160	0.2964
T7	0.0111	0.9107	0.5107	0.3809	F13	0.0152	0.9713	0.6083	0.4206
T8	0.0111	0.9110	0.5112	0.3811	TF1	0.0356	0.9939	0.6055	0.4370
T9	0.0010	0.6239	0.3618	0.2600	TF2	0.0030	0.9205	0.1400	0.3057
T10	0.0233	0.9240	0.3962	0.3681	TF3	0.0233	0.9197	0.6577	0.4191
T11	0.0050	0.9919	0.6207	0.4242	TF4	0.0111	0.9286	0.4078	0.3657
T12	0.0111	0.9247	0.1883	0.3206	TF5	0.0071	0.8813	0.3607	0.3400
T13	0.0030	0.9211	0.3240	0.3426	TF6	0.0172	0.8879	0.7433	0.4236
T14	0.0050	0.9192	0.3784	0.3540	TF7	0.0010	0.9140	0.7462	0.4239
T15	0.0030	0.9775	0.5957	0.4139	TF8	0.0010	0.9069	0.7589	0.4243
T16	0.0132	0.9429	0.5453	0.3985	PCA	0.0050	0.5864	0.0114	0.1807
F1	0.0132	0.9634	0.5547	0.4065	LLE	0.0030	0.9999	0.8071	0.4629
F2	0.0132	0.9260	0.4199	0.3684	LLTSA	0.0010	0.5866	0.4686	0.2702
F3	0.0172	0.9885	0.6996	0.4451	SE	0.0091	0.9135	0.5620	0.3910
F4	0.0193	0.9833	0.6986	0.4443	PE	0.0091	0.9925	0.7807	0.4584
F5	0.0172	0.9814	0.7183	0.4467	DE	0.0111	0.9903	0.7028	0.4432
F6	0.0254	0.9813	0.5552	0.4181	SDE	0.0193	0.9958	0.8166	0.4717

The root mean square (RMS), kurtosis (Kurt) and entropy features can characterize the energy characteristics, impact characteristics, and complexity characteristics of vibration signals, respectively. Therefore, Figure 6 shows the normalized degradation feature curves of RMS, Kurt, SE, PE, and DE, as shown by the blue curve. In order to compare the effectiveness of adaptive state warning lines based on different degradation features, the state warning lines of the above features are given by red curves. At the same time, Figure 6 also shows the normalized feature curve and corresponding state warning line of the features with high MEI score, including LLE feature, F10 feature, and F5 feature. Kurt's feature curve and MEI indicators 0.2686 both show that Kurt has poor performance in bearing condition detection. The RMS, SE, and PE feature curves can track the fault state of the bearing, but their adaptive state warning lines (red curves) cannot detect the fault warning point until after the 560th time point, and their MEI indicators 0.425, 0.391, and 0.458 are all small. The DE, LLE, F10, and F5 feature curves can better track bearing fault state, and the corresponding state warning line can detect the fault warning point near the 540th time point, which advances the warning time by nearly 20 time points.

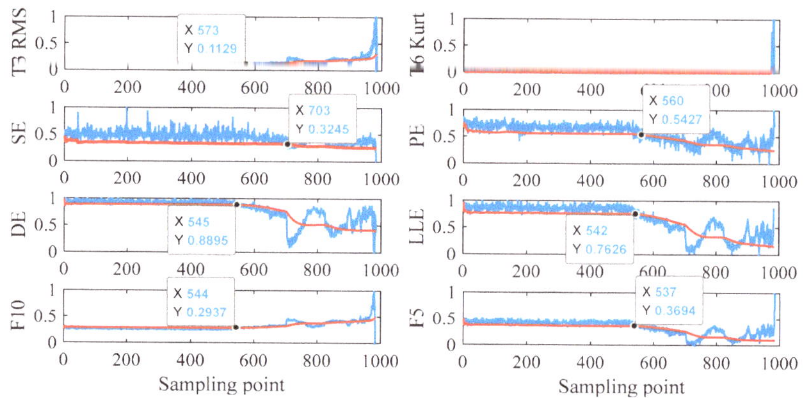

Figure 6. Bearing fault feature curve (blue curve) and state warning line (red curve).

Figure 7 shows the normalized SDE and smooth SDE feature curves and the corresponding adaptive state warning lines. With the increase of time point, the SDE feature curve experienced a change process from falling to rising, and then rising to falling, and it maintains the overall monotonicity. It can be seen from the smooth SDE feature state warning line that the bearing is in normal operation state within the time point range #1~#525, and the 525th time point is the fault early warning point. The bearing is in a slight wear state within the time point range #526~#768, the bearing is in a serious wear state within the time point range #769~#979, and the bearing is completely damaged after the 980th time point. As can be seen from the SDE feature state warning line, the 529th time point is the fault warning point and the 764th time point is the severe wear point. Compared with the fault points detected by the smooth SDE feature, the fault points detected by SDE feature are relatively lagging, which indicates that the introduction of LOWESS can suppress the influence of noise and burrs. In Table 1, the mixed evaluation index (MEI) of SDE feature is the largest, and it can be seen from Figure 7 that the 529th time point and the 525th time point are the fault warning points detected by the SDE and smooth SDE state warning lines, respectively. Compared with the single features, mixed domain fusion features and traditional entropy features, the proposed SDE feature can detect the fault warning point of rolling bearing earlier. Therefore, SDE and smooth SDE features are effective fault state detection methods, which can effectively track the state evolution process of bearing.

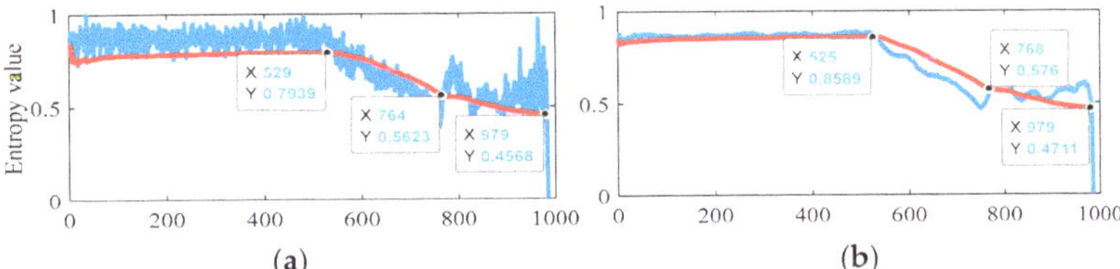

Figure 7. SDE and Smooth SDE curve (blue curve) and corresponding state warning line (red curve). (**a**) SDE; (**b**) smooth SDE.

3.2. Check Valve Fault Detection and Practical Application

The above experiments show that SDE feature can track the running state of bearing in laboratory environment more effectively. In order to verify the tracking performance of SDE to the fault state of mechanical parts in industrial environment, the proposed method is used to detect the fault degradation state of the check valve in the slurry transportation environment.

The experimental data come from the GEHO-TZPM2000 diaphragm pump (WEIR MINERALS NETHERLANDS B.V., EGTENRAYSEWEG 9 NL-5928 PH VENLO NETHER-LANDS) of the DaHongShan slurry pipeline transportation system. The maximum pressure of the main pump is 24.4 MPa, the working pressure range is 18~21 MPa, the transmission elevation difference is 1520 m and the flow rate is 350 m^2/h. The internal components of the GEHO-TZPM2000 diaphragm pump include three feed check valves and three discharge check valves. Figure 8(a1,a2) show a discharge check valve and a feed check valve respectively. The mechanical structure of the diaphragm pump check valve is shown in Figure 8b, and the spool spring forms a weakly damped oscillation system. When the diaphragm pump is running, the coordinated operation of the feed check valve and the discharge check valve makes the slurry flow smoothly. The valve core of the check valve moves back and forth in the valve chamber, the frequent contact between the valve core and the slurry often causes damage to the valve core of check valve, and Figure 8c shows a valve core that has been punctured. The sensor measuring point position and signal acquisition system are shown in Figure 9, and six PCB 352C33 sensors (PCB PIEZOTRONICS, Buffalo, NY, USA) are mounted on the shells of No. 1, No. 2, and No. 3 feed valves and No. 1, No. 2, and No. 3 discharge valves respectively. The vibration signal of the No. 1, No. 2, and No. 3 feed valves are collected through 0, 2, and 4 channels of the PXI-3342 acquisition card (Beijing Fanhua Hengxing Technology Co., Ltd, Beijing, China.), and the vibration signals of the No. 1, No. 2, and No. 3 discharge valves are collected through channels 1, 3, and 5, respectively. The sampling frequency is 2560 Hz and the data length is 76,800.

The vibration signal of the check valve is collected through a phased data acquisition scheme. The check valve is in a safe operation state within the first 500 h, so the vibration data is collected every 2 h to improve the data acquisition efficiency. The check valve may be in an early fault state from the 500th hour to the 1000th hour, so the vibration data is collected every 10 min. The check valve may be in a severe wear phase after the 1000th hour, so vibration data is collected every 2 min. Then, 421 time point samples are selected from all the time point samples obtained, and each time point sample contains 2048 data points. The sampling data is shown in Figure 10. Different from the sampling data of the bearing, the state degradation data of the check valve contains burrs and noise at any stage, and the change of amplitude is irregular. Therefore, it is more difficult to detect the fault state of the check valve compared with the bearing.

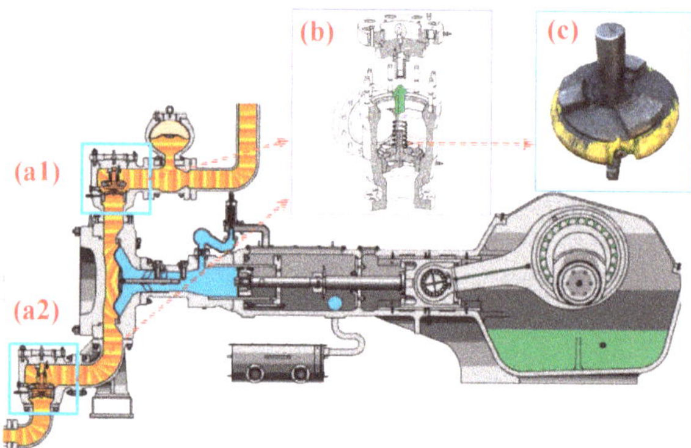

Figure 8. The structure of the diaphragm pump and check valve. (**a1**) discharge check valve; (**a2**) feed check valve; (**b**) structure of check valve; (**c**) a valve core that has been punctured.

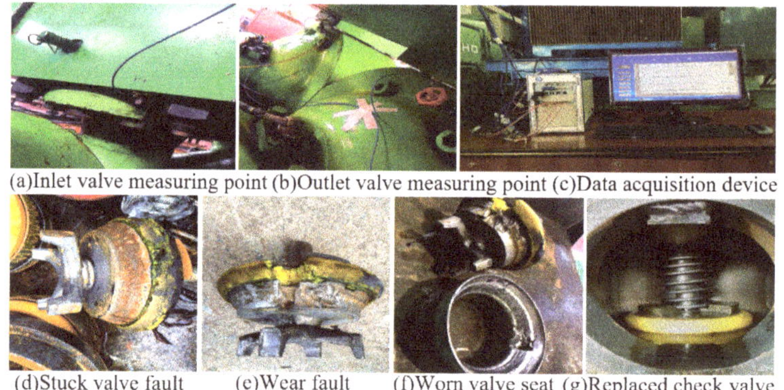

(a)Inlet valve measuring point (b)Outlet valve measuring point (c)Data acquisition device

(d)Stuck valve fault　　(e)Wear fault　　(f)Worn valve seat (g)Replaced check valve

Figure 9. Sensor measuring point, acquisition platform and fault check valve. (**a**) Inlet valve measuring point; (**b**) Outlet valve measuring point; (**c**) Data acquisition device; (**d**) Stuck valve fault; (**e**) Wear fault; (**f**) Worn valve seat; (**g**) Replaced check valve.

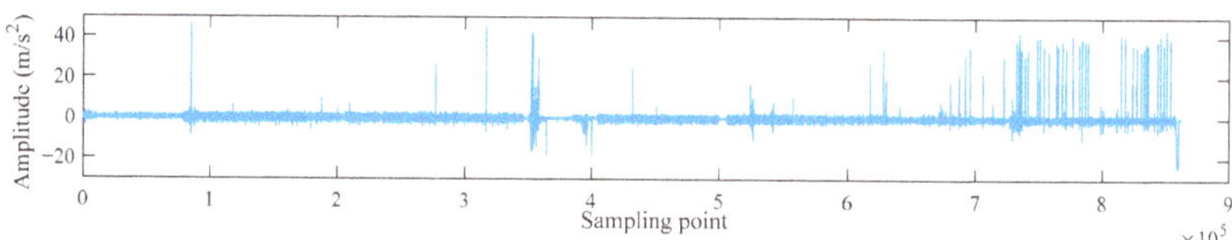

Figure 10. Check valve sampling signal.

　　In order to verify the state tracking performance of the proposed SDE feature, it is compared with time domain, frequency domain, time-frequency domain features, fusion features and entropy features. For the check valve vibration signals from the first time point to the 421th time point, the SDE features of the vibration signals at each time point are extracted respectively, and an SDE degradation curve is obtained. Among them, the first 7680 data points of the vibration signal at each time point are used to extract the fault degradation features. Similarly, for the check valve vibration signals corresponding to 421 time points, their 16 time domain features (T1~T16), 13 frequency domain features (F1~F13), 8 wavelet packet energy features (TF1~TF8), 3 mixed domain fusion features based on PCA, LLE, and LLTSA, and 3 entropy features based on SE, PE, and DE can be obtained, and the state tracking performance of all the above degradation feature curves can be evaluated by the mixed evaluation index (MEI).

As shown in Table 2, the monotonicity index (Mon) 0.0666, robustness index (Rob) 0.9494 and trend index (Tre) 0.6001 of the SDE feature are close to the corresponding maximum 0.0761, 0.9999, and 0.6846, and the MEI 0.4382 of the SDE feature is the largest, which indicates that the SDE feature has the best state tracking performance. In addition, the MEIs of T11, DE, F3, and LLE features are 0.4132, 0.4125, 0.4087, and 0.4052, respectively, which indicates that these features can also achieve good degradation state characterization performance. In addition to SDE, DE, and LLE features, those features that can characterize the operating state of the bearing cannot characterize the operating state of the check valve effectively, such as RMS, SE, PE, F10, and F5. The results show that it is more difficult to track the operating state of the check valve, and the single feature is only sensitive to a specific degradation stage or a specific fault state. Although the fusion features such as PCA, LLE, and LLTSA can characterize the fault state information of machinery, these fusion features have strong relevance and redundancy. The MEI index of the 44 degradation features of the check valve is smaller than that of bearing, which indicates that the degradation process of check valve is more complex and the state tracking is more difficult. The reason is that the vibration signal of the check valve is affected by the slurry scouring and multi-part vibration and has nonlinear and non-stationary characteristics, and the traditional degradation features are not suitable for the fault state monitoring of the check valve.

Table 2. The evaluation results of check valve fault state detection.

Feature	Mon	Rob	Tre	MEI	Feature	Mon	Rob	Tre	MEI
T1	0.0380	0.8520	0.0652	0.2877	F7	0.0571	0.9479	0.3254	0.3780
T2	0.0285	0.8501	0.0620	0.2817	F8	0.0333	0.9745	0.3589	0.3808
T3 RMS	0.0142	0.0411	0.1026	0.2800	F9	0.0333	0.9621	0.2066	0.3466
T4	0.0380	0.8520	0.0652	0.2877	F10	0.0380	0.8982	0.3893	0.3663
T5	0.0428	0.6580	0.0841	0.2356	F11	0.0095	0.8498	0.3105	0.3218
T6 Kurt	0.0571	0.5694	0.0840	0.2162	F12	0.0190	0.7823	0.1837	0.2809
T7	0.0095	0.7583	0.5998	0.3522	F13	0.0047	0.8443	0.3905	0.3338
T8	0.0142	0.7584	0.5962	0.3539	TF1	0.0285	0.7802	0.6846	0.3952
T9	0.0190	0.6151	0.0824	0.2105	TF2	0.0238	0.7716	0.6425	0.3719
T10	0.0142	0.7518	0.0836	0.2494	TF3	0.0047	0.7035	0.4003	0.2935
T11	0.0047	0.9547	0.6221	0.4132	TF4	0.0047	0.7427	0.5321	0.3316
T12	0.0380	0.7902	0.6644	0.3889	TF5	0.0428	0.6275	0.4898	0.3076
T13	0.0095	0.7778	0.6722	0.3725	TF6	0.0047	0.6974	0.4431	0.3002
T14	0.0047	0.7746	0.6716	0.3691	TF7	0.0190	0.7107	0.3904	0.3008
T15	0.0142	0.8359	0.6561	0.3891	TF8	0.0047	0.7117	0.4510	0.3061
T16	0.0190	0.7838	0.6324	0.3711	PCA	0.0428	0.9920	0.0840	0.3358
F1	0.0142	0.8139	0.3309	0.3175	LLE	0.0071	0.9999	0.5083	0.4052
F2	0.0190	0.7601	0.0829	0.2541	LLTSA	0.0095	0.6520	0.1708	0.2345
F3	0.0047	0.9328	0.6328	0.4087	SE	0.0095	0.8861	0.3336	0.3373
F4	0.0142	0.9112	0.6350	0.4055	PE	0.0238	0.9742	0.4658	0.3973
F5	0.0761	0.9338	0.3794	0.3941	DE	0.0285	0.9581	0.5538	0.4125
F6	0.0285	0.8726	0.4408	0.3642	SDE	0.0666	0.9494	0.6001	0.4382

The RMS, Kurt, SE, PE, and DE features and their state warning lines are shown in Figure 11. In addition, Figure 11 also shows the T11, F3 and LLE feature curves with large MEI indexes. The RMS (MEI = 0.28), Kurt (MEI = 0.216), SE (MEI = 0.337), and PE (MEI = 0.397) features that can characterize the degradation state of the bearing cannot characterize the degradation state of the check valve, and these features can only detect the abnormal state at the 178th or 179th time point. The feature curves of DE, T11, F3, and LLE contain many burrs, and the phased variation of the curve is not obvious. In addition, the state warning line of these four features contains several turning points. Based on this, it can be determined that the fault warning point is roughly between the 170th time point and the 178th time point, and the check valve is in the severe wear phase between the 330th time point and the 421st time point. This is due to the complex structure of the diaphragm pump and the variability of influencing factors in the slurry transportation environment, which makes it difficult to track the fault state of the check valve.

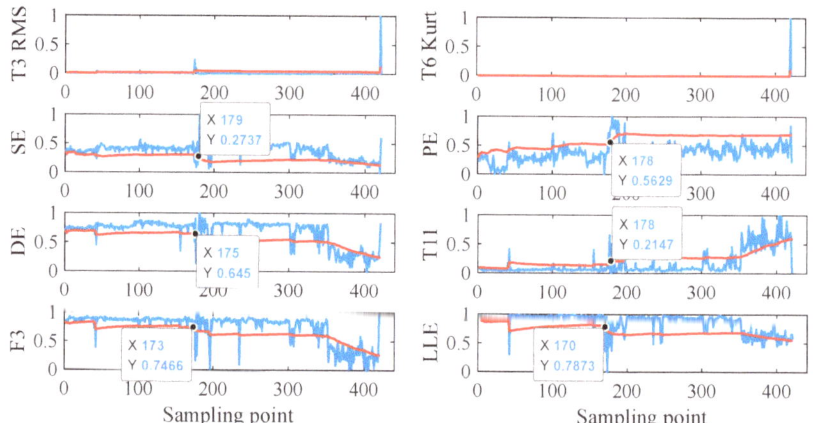

Figure 11. Check valve fault feature curve (blue curve) and status warning line (red curve).

The fault features of the check valve are not obvious in the slurry transportation environment. If the damage is not detected until it develops to a certain extent, the repair time will be insufficient and the repair cost may increase exponentially. Figure 12 shows the normalized SDE feature and smooth SDE feature of the check valve, and the corresponding state warning lines (red curve). As shown in Figure 12a, the SDE feature is close to 1 and the check valve is in a normal state from the first time point to the 169th time point. The SDE feature has a significant decline in stages from the 170th time point to the 326th time point, and the check valve may be in a state of slight wear. The decline of the SDE feature is more obvious from the 327th time point to the 421st time point, and the check valve may be in a state of severe wear. However, the SDE feature curve contains many burrs and fluctuations, which makes it difficult to determine the fault warning point through the state warning line, and the SDE feature can only detect the 170th time point as the state mutation point. Therefore, it is particularly necessary to track the state of check valve through the proposed smooth SDE. As shown in Figure 12b, the state warning line of smooth SDE can detect that the 168th time point is the fault warning point, and the 318th time point is the key state point. In addition, the smooth SDE feature can track the operation state of the check valve effectively. The tracking results show that the check valve is in a normal state from the first time point to the 168th time point, slightly worn from the 169th time point to the 318th time point, and seriously worn from the 319th time point to the 421st time point. Under the influence of noise and burrs, the smooth SDE feature and its state warning line can track the degradation state of the check valve effectively and detect the fault warning point earlier.

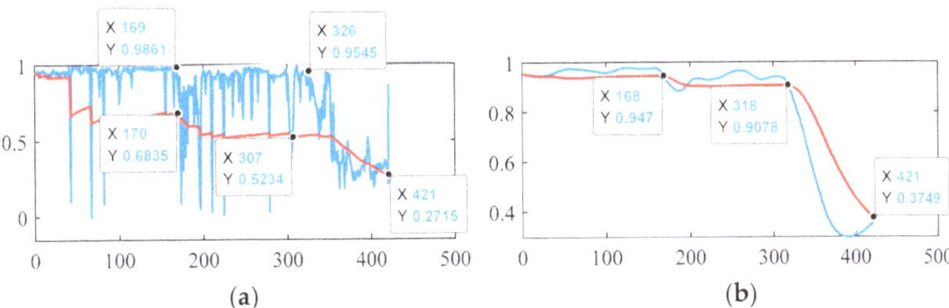

Figure 12. SDE and SDE feature curve (blue curve) and corresponding state warning line (red curve). (**a**) SDE; (**b**) smooth SDE.

3.3. Comparison and Discussion

Through the two sets of experiments, the effectiveness and superiority of the proposed SDE and its adaptive state warning line have been proved. In order to further verify the superiority of the method proposed in this paper, the proposed SDE is also compared with the latest research results. Under the background of the same IMS bearing degradation experiment, the SDE proposed in this paper (Figure 7) can detect the fault warning point adaptively at the 525th time point. However, the energy moment entropy [22] proposed by Gao cannot detect the fault warning point until the 538th time point. Similarly, the oscillation based permutation entropy [13] proposed by Noman cannot detect the fault warning point until the 533th time point. The results show that the SDE proposed in this paper can detect the fault warning point of mechanical parts earlier and provide more response time for the repair and replacement of parts. Although the complex wavelet packet energy moment entropy [23] proposed by Shao can also detect fault warning point at the 525th time point, the determination of the fault warning point depends on manual experience and lacks an adaptive state warning line. In contrast, the method proposed in this paper can track and detect early fault warning points adaptively. In addition, compared with the latest research results, the proposed SDE has a high tracking efficiency, and the smoothness and effectiveness of the SDE tracking curve are significant. To sum up, the proposed fault state detection method based on SDE has outstanding effectiveness and superiority.

4. Conclusions

In this paper, a fault state detection and evaluation method based on SDE is proposed, which can track the degradation state of bearing and check valve and detect the operation state of mechanical parts at the current time. Through the analysis of the IMS bearing data in the laboratory environment and the check valve data in the industrial environment, the effectiveness of the proposed method is proved. By comparing the proposed SDE with single features, fusion feature and traditional entropy feature, the following conclusions can be drawn.

(1) In the condition monitoring of check valve and bearing, the MEI scores of SDE features are 0.4382 and 0.4717, respectively, and these two scores are much higher than those of single features, fusion features, and traditional entropy features. The results show that sliding window down-sampling improves the trend of degradation features, TANSIG mapping enhances the performance of SDE features to characterize degradation states, and the introduction of LOWESS improves the anti-interference performance of features. The SDE feature and its state warning line can effectively track the operation state of the check valve and determine the fault warning point earlier.

(2) The reason why the MEI scores of the 44 degradation features of the check valve is smaller than that of the 44 degradation features of the bearing is that the vibration signal of check valve in industrial environment is affected by factors such as slurry erosion

and multi-part vibration. Even so, the proposed smooth SDE feature can still detect the degradation state of the check valve effectively.

(3) A new method for fault detection of mechanical parts is proposed in this paper, which can not only guide the formulation of maintenance and replacement plan, but also improve the operation safety of diaphragm pump and other equipment. Next, we will study the fault trend prediction methods and early fault diagnosis techniques.

Author Contributions: Conceptualization, C.Z. and Y.J.; methodology, C.Z.; software, Y.J.; validation, H.B., Y.Y. and L.X.; formal analysis, C.Z.; investigation, C.Z.; resources, Y.Y.; data curation, C.Z.; writing—original draft preparation, H.B.; writing—review and editing, C.Z.; visualization, C.Z.; supervision, H.B. and Y.Y.; project administration, Y.Y.; funding acquisition, Y.Y. All authors have read and agreed to the published version of the manuscript.

Funding: This research was funded by the National Natural Science Foundation of China Program (No.61663017 and No. 41971392), PhD research startup foundation of Yunnan Normal University (No.01000205020503131) and Yunnan Province Ten-thousand Talents Program.

Institutional Review Board Statement: Not applicable.

Informed Consent Statement: Informed consent was obtained from all subjects involved in the study.

Data Availability Statement: The data used to support the findings of this study are available from the corresponding author upon request.

Acknowledgments: This work was supported by the National Natural Science Foundation of China (No.61663017 and No. 41971392) and Yunnan Province Ten-thousand Talents Program. The author sincerely thanks the team for their guidance, and thanks the Case West Reserve University and Yunnan Dahongshan pipeline company for their bearing datasets and check valve datasets. The author sincerely expresses thanks to the reviewers for taking the time to review the paper in a busy schedule.

Conflicts of Interest: The authors declare no conflict of interest.

References

1. Li, W.; Mckeown, A.; Yu, Z. Correction of cavitation with thermodynamic effect for a diaphragm pump in organic Rankine cycle systems. *Energy Rep.* **2020**, *6*, 2956–2972. [CrossRef]
2. Xiao, L.; Liu, Z.; Zhang, Y.; Zheng, Y.; Cheng, C. Degradation assessment of bearings with trend-reconstruct-based features selection and gated recurrent unit network. *Measurement* **2020**, *165*, 108064. [CrossRef]
3. Gao, S.; Zhang, S.; Zhang, Y.; Gao, Y. Operational reliability evaluation and prediction of rolling bearing based on isometric mapping and NoCuSa-LSSVM. *Reliab. Eng. Syst. Saf.* **2020**, *201*, 106968. [CrossRef]
4. Hua, Z.; Xiao, Y.; Cao, J. Misalignment Fault Prediction of Wind Turbines Based on Improved Artificial Fish Swarm Algorithm. *Entropy* **2021**, *23*, 692. [CrossRef] [PubMed]
5. Li, Z.; Zhang, X.; Kari, T.; Hu, W. Health Assessment and Remaining Useful Life Prediction of Wind Turbine High-Speed Shaft Bearings. *Energies* **2021**, *14*, 4612. [CrossRef]
6. Bilendo, F.; Badihi, H.; Lu, N.; Jiang, B. A data-driven prognostics method for explicit health index assessment and improved remaining useful life prediction of bearings. *ISA Trans.* **2021**. [CrossRef] [PubMed]
7. He, M.; Zhou, Y.; Li, Y.; Wu, G.; Tang, G. Long short-term memory network with multi-resolution singular value decomposition for prediction of bearing performance degradation. *Measurement* **2020**, *156*, 107582. [CrossRef]
8. Ding, N.; Li, H.; Yin, Z.; Jiang, F. A novel method for journal bearing degradation evaluation and remaining useful life prediction under different working conditions. *Measurement* **2021**, *177*, 109273. [CrossRef]
9. Hu, M.; Wang, G.; Ma, K.; Cao, Z.; Yang, S. Bearing performance degradation assessment based on optimized EWT and CNN. *Measurement* **2021**, *172*, 108868. [CrossRef]
10. Shannon, C.E. A mathematical theory of communication. *Mob. Comput. Commun. Rev.* **2001**, *5*, 3–55. [CrossRef]
11. Qin, A.-S.; Mao, H.-L.; Hu, Q. Cross-domain fault diagnosis of rolling bearing using similar features-based transfer approach. *Measurement* **2021**, *172*, 108900. [CrossRef]
12. Kumar, P.S.; Kumaraswamidhas, L.; Laha, S. Selection of efficient degradation features for rolling element bearing prognosis using Gaussian Process Regression method. *ISA Trans.* **2021**, *112*, 386–401. [CrossRef] [PubMed]
13. Noman, K.; Wang, D.; Peng, Z.; He, Q. Oscillation based permutation entropy calculation as a dynamic nonlinear feature for health monitoring of rolling element bearing. *Measurement* **2021**, *172*, 108891. [CrossRef]

14. Minhas, A.S.; Kankar, P.; Kumar, N.; Singh, S. Bearing fault detection and recognition methodology based on weighted multiscale entropy approach. *Mech. Syst. Signal Process.* **2021**, *147*, 107073. [CrossRef]
15. Li, H.; Sun, J.; Ma, H.; Tian, Z.; Li, Y. A novel method based upon modified composite spectrum and relative entropy for degradation feature extraction of hydraulic pump. *Mech. Syst. Signal Process.* **2019**, *114*, 399–412. [CrossRef]
16. Mostafa, R.; Reza, A.M.; Hamed, A. Application of dispersion entropy to status characterization of rotary machines. *J. Sound Vib.* **2018**, *438*, 291–308.
17. Rostaghi, M.; Azami, H. Dispersion Entropy: A Measure for Time-Series Analysis. *IEEE Signal Process. Lett.* **2016**, *23*, 610–614. [CrossRef]
18. Zhang, B.; Zhang, S.; Li, W. Bearing performance degradation assessment using long short-term memory recurrent network. *Comput. Ind.* **2019**, *106*, 14–29. [CrossRef]
19. Azami, H.; Rostaghi, M.; Escudero, J. Refined composite multiscale dispersion entropy: A fast measure of complexity. *IEEE Trans. Bio-Med. Eng.* **2016**, *99*, 1–5.
20. Cleveland, W.S. LOWESS: A Program for Smoothing Scatterplots by Robust Locally Weighted Regression. *Am. Stat.* **1981**, *35*, 54. [CrossRef]
21. Duong, P.B.; Khan, A.S.; Shon, D.; Im, K. A Reliable Health Indicator for Fault Prognosis of Bearings. *Sensors* **2018**, *18*, 3740. [CrossRef] [PubMed]
22. Gao, Z.; Liu, Y.; Wang, Q.; Wang, J.; Luo, Y. Ensemble empirical mode decomposition energy moment entropy and enhanced long short-term memory for early fault prediction of bearing. *Measurement* **2021**, 110417. [CrossRef]
23. Haidong, S.; Junsheng, C.; Hongkai, J.; Yu, Y.; Zhantao, W. Enhanced deep gated recurrent unit and complex wavelet packet energy moment entropy for early fault prognosis of bearing. *Knowl. Based Syst.* **2020**, *188*, 105022. [CrossRef]

Article

Dynamic Modeling and Simulation Analysis of Inter-Shaft Bearings with Local Defects Considering Elasto-Hydrodynamic Lubrication

Jing Tian, Xinping Ai, Fengling Zhang *, Zhi Wang, Cai Wang and Yingtao Chen

Liaoning Key Laboratory of Advanced Test Technology for Aeronautical Propulsion System, Shenyang Aerospace University, Shenyang 110136, China
* Correspondence: fling707@sau.edu.cn; Tel.: +86-18842471309

Abstract: As an important component of large engines, inter-shaft bearing is easily damaged due to its poor working conditions. By analyzing the time–frequency distribution rules of fault signals and the evolution law of micro-faults, the bearing failure mechanism can be revealed, and the bearing failure can be monitored in real time and prevented in advance. For the purpose of studying the mechanism of inter-shaft bearing faults, a 4-DOF (degree of freedom) dynamic model of inter-shaft bearing with local defects considering elasto-hydrodynamic lubrication (EHL) is proposed. Based on the established dynamic model, the impact characteristics and distribution rules of the fault signals of the bearing are accurately simulated, and the evolution law of the micro-faults is also analyzed. The effects of different speeds, loads and defect widths on maximum value (MV), absolute mean value (AMV), effective value (EV), amplitude of square root (AST), kurtosis factor (KF), impulse factor (IF), peak factor (PF) and shape factor (SF) are obtained. The findings show that the vibration amplitude of the bearing increases with the increase in defect size, and the faults are easier to diagnose accordingly. At the same time, PF, KF and IF are very sensitive to the initial failure of bearings. With the development of faults, the overall trend of AMV, AST and EV are relatively stable. The PF is sensitive to the change of rotating speeds and defect widths. The SF is insensitive to the change of rotating speeds, loads and defect widths. This lays a foundation for the research of monitoring and diagnosis methods of aeroengine inter-shaft bearing fault.

Keywords: inter-shaft bearings; local defects; time-varying displacement; elasto-hydrodynamic lubrication; defect widths

Citation: Tian, J.; Ai, X.; Zhang, F.; Wang, Z.; Wang, C.; Chen, Y. Dynamic Modeling and Simulation Analysis of Inter-Shaft Bearings with Local Defects Considering Elasto-Hydrodynamic Lubrication. *Coatings* **2022**, *12*, 1735. https://doi.org/10.3390/coatings12111735

Academic Editors: Andrew J. Pinkerton and Georgios Skordaris

Received: 30 September 2022
Accepted: 8 November 2022
Published: 13 November 2022

Publisher's Note: MDPI stays neutral with regard to jurisdictional claims in published maps and institutional affiliations.

1. Introduction

Inter-shaft bearings are widely used in the supporting drive systems of twin-rotor aero engines. They are one of the necessary parts of the areoengine-supporting drive system. Due to the harsh working conditions of aeroengine, faults often occur in inter-shaft bearings.

At present, the dynamic modeling method of rolling bearing with local defects is widely studied. Rolling elements are equivalent to nonlinear springs, and a 2-DOF model is mainly for the dynamic analysis of a transient state during the running of the rolling bearing [1]. At the same time, Gupta [2–9] published many models to describe motion states and force states of each part and to consider the speed changes of each part and the corresponding effect of inertial force. However, the model does not consider the influence of damping, and it only simplifies the collision contact according to the elastic contact. Based on the McFadden model [10], Su [11,12] established the dynamic model of bearing with defects and distribution defects under variable loads and used this model to reveal the frequency characteristics of the two defects. Hu [13] proposed a 5-DOF dynamic model of deep-groove ball bearings. The model theoretically formulated the elastic deformation and nonlinear contact forces of bearings coupling dual rotors. Patel [14] established a 6-DOF dynamic model for deep-groove ball bearings and calculated the vibration response of single-point and multi-point faults under constant load. Based on the Hertz contact theory, Xi [15], Ma [16], Patel [17], Liu [18] and Cao [19] established bearing dynamic models with different degrees of freedom and analyzed the nonlinear response of rolling bearing with local defects.

Patil [20] considered the rolling element as a nonlinear contact spring and proposed a dynamic model to research the effects of the defect size on the vibration characteristics of the bearing. Based on the Patil model, Kulkarni [21] used cubic Hermite spline interpolation difference functions to simulate the pulse generated by bearing faults. Cui [22] established a nonlinear dynamic model, which used rectangular displacement excitation to simulate local defects. Khanam [23] proposed a dynamic model using a semi-sinusoidal curve to describe the displacement excitation. Wang [24] came up with a multi-body dynamic model to study the vibration response of cylindrical roller bearings with local defects and analyzed the influence of time-varying contact on bearing defects. The above research shows that the dynamic model of rolling bearing based on the Hertz contact theory, combined with the displacement excitation function, can accurately simulate the vibration signals generated by bearing surface faults.

Sassi [25] established a 3 DOF bearing vibration equation considering the influence of an oil film. The interaction between the rollers and the fault area were analyzed. Wang [26] developed a coupling model of the rotor bearings system, considering the oil film. The variation laws of oil films were analyzed based on this model. Yan [27] and Zhang [28] also considered the influence of the lubricating oil film of the bearing in the established dynamic model and proved that it can more accurately describe the real-time state of bearing operation when considering the influence of EHL on the bearing.

In addition to establishing models for rolling bearing with local defects, the researchers also proposed a variety of coupling modeling methods for fault bearing and rotor systems [29–32]. Niu [33] established a dynamic model considering the change in contact force direction of roller sliding and entering defects. Cao [34] proposed a dynamic model method of bearing, considering a bearing pedestal and rotor system. To sum up, research on the dynamic model of rolling bearing fault carried out by experts and scholars mostly focuses on the faults of conventional rolling bearing. There is not much research on the modeling of inter-shaft bearing fault dynamics. Rolling bearing and inter-shaft bearing have both similarities and differences. The modeling method of rolling bearing can be used as the basis of inter-shaft bearing modeling, but it can not completely describe the motion state of inter-shaft bearing. In this paper, the inter-shaft bearing is taken as the main research object. Based on the nonlinear Hertz contact theory, a local defect dynamic model of the inter-shaft bearing, considering the influence of time-varying displacement excitation (TVDE) and elasto-hydrodynamic lubrication (EHL), is proposed. The model is used to simulate and analyze the time–frequency distribution rules of fault signals and the evolution law of micro-faults.

Other parts of this paper are arranged as follow. The fault simulation experiment of co-rotating and counter rotating on the birotor bearing test rig is shown in Part 3. The fault characteristics of inter-shaft bearing with local defects are studied in Part 4, and the time–frequency characteristics of micro faults are also studied in this section. Finally, part 5 summarizes the conclusion of this paper.

2. Establishment of Fault Dynamic Model

2.1. Subsection Simplification and Assumption of Inter-Shaft Bearings

Inter-shaft bearing is an important component of aeroengine rotor systems and works between an HP (high-pressure) rotor and an LP (low-pressure) rotor. Unlike ordinary bearings, the inner ring (IR) and outer ring (OR) of bearings rotate at the same time. Depending on the rotation direction, the working speed of the bearing is the speed difference or sum of the two rotors speed. The supporting form of typical dual-rotor aeroengine inter-shaft bearing is shown in Figure 1.

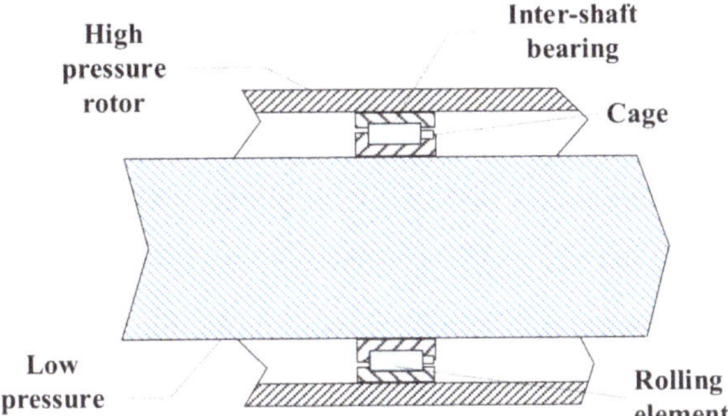

Figure 1. Supporting form of inter-shaft bearing.

The IR and OR of the inter-shaft bearing rotate at the same time, so it is considered to generate vibration frequently during operation. Considering that both the IR and OR have vertical and horizontal vibrations, a 4-DOF dynamic model is established in this paper. The model assumes that the rollers do not slip. Referring to the Patil model [20], the contact between rollers and rings is simplified as a nonlinear spring-mass system. The model is shown in Figure 2.

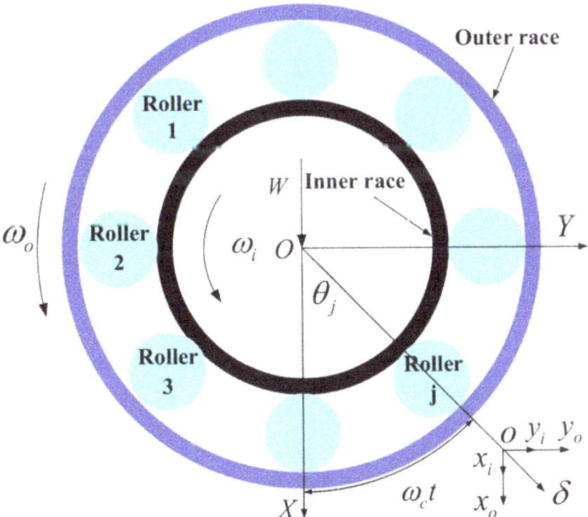

Figure 2. Simplified model of inter-shaft bearing.

2.2. Nonlinear Hertz Contact Forces

On the basis of the nonlinear Hertz contact theory [35], Harris [36] derived the relationship between the nonlinear load and the displacement of the rolling bearing as shown in Equation (1).

$$F = K\delta^n \tag{1}$$

where K is the load deformation coefficient. δ is the radial displacement. Cylindrical roller bearing $n = 10/9$.

Since the rollers in the inter-shaft bearing has contact with both the IR and OR, the total contact stiffness is as follows.

$$K = \left[\frac{1}{(1/K_i)^{1/n} + (1/K_0)^{1/n}} \right]^n \tag{2}$$

where K_i is the contact stiffness of the IR. K_0 is the contact stiffness of the OR. For cylindrical roller bearing, $K = 8.06 \times 10^4 \times l^{8/9}$, where l is the roller length.

δ_j, the radial deformation of the j roller is

$$\delta_j = (x_0 - x_i)\cos\theta_j + (y_0 - y_i)\sin\theta_j - C_r - H_d \tag{3}$$

where x_0 and y_0 are the displacement of the center of the OR along the x-axis and y-axis. x_i and y_i is the displacement of the center of the IR along the x-axis and the y-axis. θ_j is the angle between the center of the j_{th} roller and the x-axis. C_r is the initial clearance between the roller and the raceway. H_d is the TVDE when the roller passes through the defect.

$$\theta_j = \omega_c t + \frac{2\pi(j-1)}{Z} + \theta_{t0} \tag{4}$$

$$\omega_c = \frac{1}{2}\left[\omega_i\left(1 - \frac{d}{D_m}\cos\alpha\right) + \omega_o\left(1 + \frac{d}{D_m}\cos\alpha\right)\right] \tag{5}$$

where ω_i is the angular velocity of IR of inter-shaft bearing, ω_o is the angular velocity of OR of inter-shaft bearing, ω_c is the angular velocity of inter-shaft bearing retainer, θ_{t0} is the initial Angle of the first roller relative to the axis, Z is the number of roller of inter-shaft bearing, d is the roller diameter, D_m is the pitch diameter of inter-shaft bearing, and α is the bearing pressure angle.

Substitute Equations (2) and (3) into Equation (2) to obtain the nonlinear Hertz contact force of the j roller:

$$F_j = K\left[(x_0 - x_i)\cos\theta_j + (y_0 - y_i)\sin\theta_j - C_r - H_d\right]^{10/9} \tag{6}$$

The total nonlinear Hertz contact force received by the inter-shaft bearing is the sum of the nonlinear Hertz contact forces of all the rollers. The components of the total contact force on the x and y axes are:

$$\begin{cases} F_X = K\sum_{j=1}^{Z}\left[(x_0 - x_i)\cos\theta_j + (y_0 - y_i)\sin\theta_j - C_r - H_d\right]^{10/9}\cos\theta_j \\ F_Y = K\sum_{j=1}^{Z}\left[(x_0 - x_i)\cos\theta_j + (y_0 - y_i)\sin\theta_j - C_r - H_d\right]^{10/9}\sin\theta_j \end{cases} \tag{7}$$

2.3. Time-Varying Displacement Excitation

This paper assumes that the defect of the inter-shaft bearing runs through the entire raceway. Therefore, the defect length is larger than the roller, but the width of the fault is far less than the diameter of the roller. When the roller passes through the defect of the raceway, the roller will not contact the bottom of the defect, and the falling displacement of the roller center will be less than the defect depth.

The relationship between and bearing movement is shown in Figure 3. In the figure, H_d is the time-varying displacement when the roller passes through the defect, H is the defect depth, B is width of the defect, and H_e is the maximum deviation of the roller center when the roller passes through the defect. It can be seen from Figure 4 that the roller can be divided into three stages when it passes through the defect. The first stage is that the roller starts to enter the defect and completely enters the defect. At this time, the time-varying displacement H_d increases gradually as the roller enters the defect; in the meantime, the roller always contacts the left side of the defect. In stage 2, the roller completely enters the defect. At this time, the roller contacts the left and right sides of the defect, and H_d reaches the peak value of H_e. In stage 3, the roller enters the defect completely and leaves the defect just after. At this time, the roller only contacts the right side of the defect, and H_d gradually decreases. To sum up, while the roller passes through the defect, H_d increases first and then decreases with motion process. Therefore, the semi-sinusoidal function is used to describe the TVDE of the local defect to the roller. The maximum displacement excitation H_e of the roller is:

$$H_e = d/2 - \sqrt{(d/2)^2 - (B/2)^2} \tag{8}$$

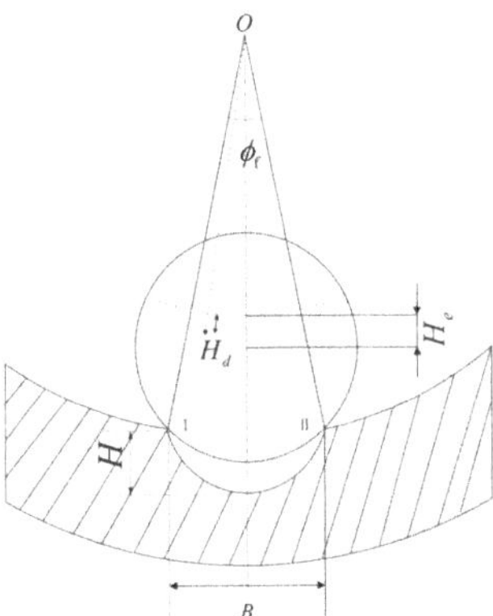

Figure 3. TVDE of the roller.

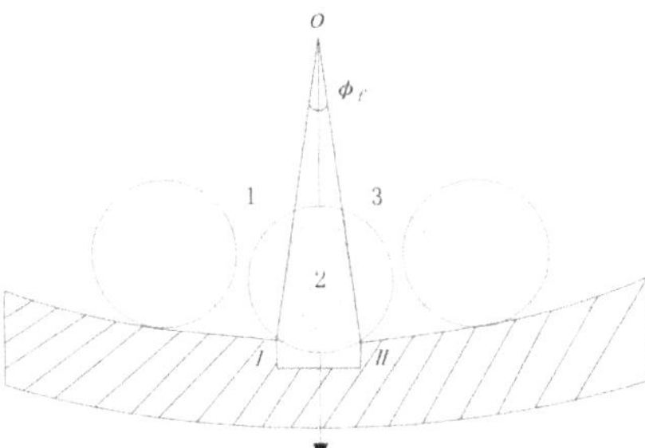

Figure 4. Roller passing defect state diagram.

2.3.1. The TVDE of Defects in the OR

Figure 5 reveals the relationship between the rollers and the defect angle position when the IR and OR of the bearing rotation in the opposite direction. The angular velocity of the IR w_i is clockwise, and the angular velocity of the OR w_0 is counterclockwise. When w_i is smaller than w_0, the angle velocity of the cage w_c is counterclockwise at this time, and φ_f is the defect angle. The counterclockwise direction is specified in the figure as the positive direction. θ_{b0} is the angle of the first roller relative to the X-axis at the initial moment. At moment t, θ_{bi} is the rotational angle of the i_{th} roller. θ_{f0} is the angle relative to the X-axis at the initial moment of the defect. θ_f is the rotational angle of the defect. The TVDE H_d generated when the roller passes through the defect is:

$$H_d = \begin{cases} H_e \sin\left(\frac{\pi}{\varphi_f}(\mathrm{mod}(\theta_f, 2\pi) - \mathrm{mod}(\theta_{bi}, 2\pi))\right) \\ \quad 0 \le \mathrm{mod}(\theta_f, 2\pi) - \mathrm{mod}(\theta_{bi}, 2\pi) \le \varphi_f \\ \\ 0 \qquad\qquad\qquad else \end{cases} \tag{9}$$

where mod is the remainders. The expressions for θ_{bi} and θ_f are:

$$\theta_{bi} = \frac{2\pi}{Z}(i-1) + w_c t + \theta_{b0}(i = 1 \ldots Z) \tag{10}$$

$$\theta f = w_o t + \theta_{f0} \tag{11}$$

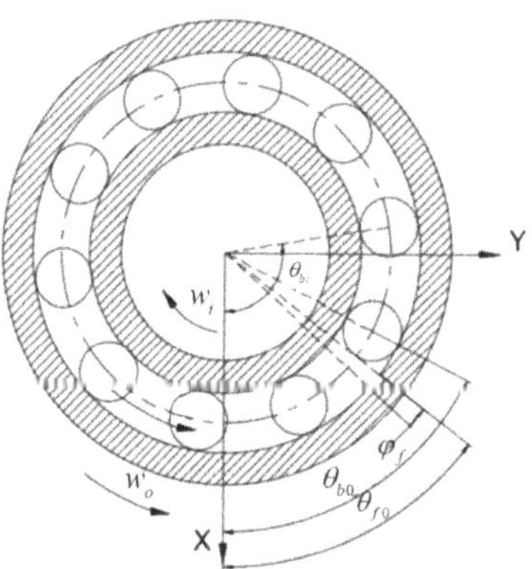

Figure 5. Roller and defect angular position of OR fault.

2.3.2. The TVDE with Defects in the IR

Figure 6 shows the relationship of rollers and defects when the IR and OR rotate in the opposite direction. When w_i is smaller than w_0, both w_c and w_0 are in the same direction. Counterclockwise is the positive direction. The expression of TVDE H_d is:

$$H_d = \begin{cases} H_e \sin\left(\frac{\pi}{\varphi_f}\left(\mathrm{mod}(\theta_{bi}, 2\pi) - \mathrm{mod}\left(\theta_{f0} + 2\pi - \mathrm{mod}(w_i t, 2\pi), 2\pi\right)\right)\right) \\ 0 \leq \mathrm{mod}(\theta_{bi}, 2\pi) - \mathrm{mod}\left(\theta_{f0} + 2\pi - \mathrm{mod}(w_i t, 2\pi), 2\pi\right) \leq \varphi_f \\ 0 \qquad\qquad\qquad\qquad\qquad\qquad else \end{cases} \tag{12}$$

$$\theta_{bi} = \frac{2\pi}{Z}(i-1) + w_c t + \theta_{b0}(i = 1 \ldots Z) \tag{13}$$

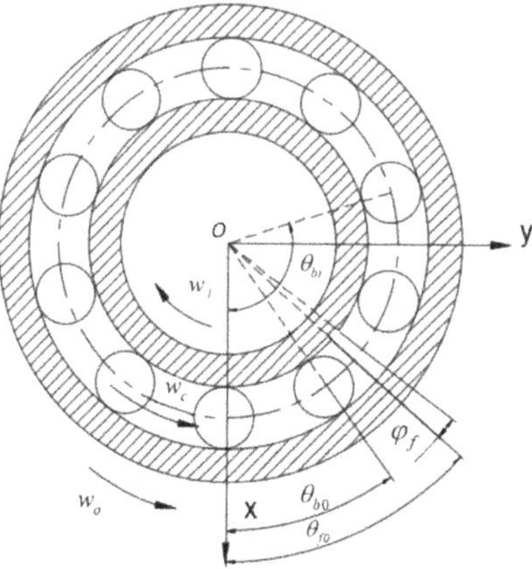

Figure 6. Roller and defect angular position of IR fault.

2.3.3. TVDE of Rolling Element Fault

Figure 7 shows the relationship between rollers and the defect angle position when the IR and OR rotate in reverse. Set w_i to be clockwise rotation and w_0 to be counterclockwise rotation. When w_0 is greater than w_i, w_c is counterclockwise, and the angle velocity of the roller w_b is counterclockwise. The counterclockwise direction is the positive direction. Assume the roller 1 has a local defect fault. The defect angle is φ_{bf}. rack the motion of the first roller and establish another moving coordinate system on axis om. The counterclockwise direction is taken as the positive direction, and θ_{bf0} is the initial angle of right defect relative to the om-axis. At moment t, the angle of the relative moving om-axis is θ_{bf}, and the TVDE H_d of the bearing is:

$$H_d = \begin{cases} H_e \sin\left(\frac{\pi}{\varphi_{bf}}\mathrm{mod}(2\pi + w_b t - \theta_{bf0}, 2\pi)\right) \\ 0 \leq \mathrm{mod}(2\pi + w_b t - \theta_{bf0}, 2\pi) \leq \varphi_{bf} \\ H_e \sin\left(\frac{\pi}{\varphi_{bf}}(\mathrm{mod}(2\pi + w_b t - \theta_{bf0}, 2\pi) - \pi)\right) \\ \pi \leq \mathrm{mod}(2\pi + w_b t - \theta_{bf0}, 2\pi) \leq \pi + \varphi_{bf} \\ 0 \qquad\qquad\qquad else \end{cases} \tag{14}$$

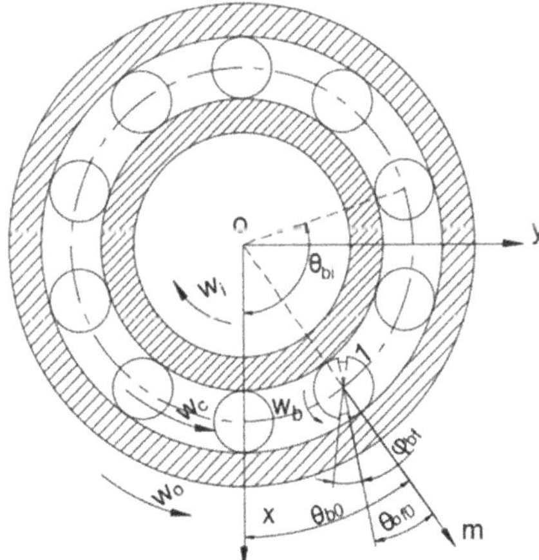

Figure 7. Roller and defect angular position of roller fault.

2.4. Dynamic Model for Inter-Shaft Bearings with Local Defects

Based on the hypothesis theory of rigid rings and considering the eccentric load on the bearing, the inter-shaft bearing is established as a 4-DOF dynamic model. The contact stiffness, damping, nonlinear Hertz contact force, eccentric load, time-varying displacement and constant radial load are substituted into the dynamic equation to establish the dynamic model of an inter-shaft bearing fault.

$$\begin{cases} M_0 \ddot{x}_0 + c\dot{x}_0 + K\sum_{j=1}^{Z} \lambda\delta^{10/9}\cos\theta_j = W + F_{l0}\cos(\omega_0 t) \\ M_0 \ddot{y}_0 + c\dot{y}_0 + K\sum_{j=1}^{Z} \lambda\delta^{10/9}\sin\theta_j = F_{l0}\sin(\omega_0 t) \\ M_i \ddot{x}_i + c\dot{x}_i - K\sum_{j=1}^{Z} \lambda\delta^{10/9}\cos\theta_j = W + F_{li}\cos(\omega_i t) \\ M_i \ddot{y}_i + c\dot{y}_i - K\sum_{i=1}^{Z} \lambda\delta^{10/9}\sin\theta_j = F_{li}\sin(\omega_i t) \end{cases} \tag{15}$$

where M_i and M_0 is the quality of IR and OR; F_{li} is the eccentric load applied to the IR; c is the damping coefficient; F_{l0} is the eccentric load applied to the OR. W is the radial load perpendicular to inner surface of IR; λ is the switch signal to indicate if the roller and the raceway are in contact, expressed as:

$$\lambda = \begin{cases} 1 & \delta > 0 \\ 0 & \delta \leq 0 \end{cases} \tag{16}$$

The dynamic differential equation is solved by the Newmark-β method [37]. Taking $\gamma = 1/2$ and $\beta = 1/4$, this method has second-order accuracy and is an unconditionally stable method, which can accurately solve the fault dynamic model of inter-shaft bearing.

2.5. Fault Dynamic Model Considering the Influence of EHL

The inter-shaft bearing is running at a high speed between high-pressure rotors and low-pressure rotors. Therefore, the inter-shaft bearing needs real-time lubrication. There is an oil film between the rollers and the raceway, and it is always in the state of EHL under the dynamic pressure. The distribution of oil-film pressure in the inter-shaft bearing is shown in Figure 8. The oil-film thickness under the EHL state affects the tribological and dynamic characteristics of the friction pair. The oil pressure generated by the high-speed rotation of the bearing makes the oil film produce a "stiffening effect". The EHL pressure curve is similar to the Hertz pressure curve in distribution. Therefore, in the dynamic model considering the influence of EHL established in this paper, the stiffness of bearing is considered as the series stiffness of oil-film stiffness and contact stiffness [37,38].

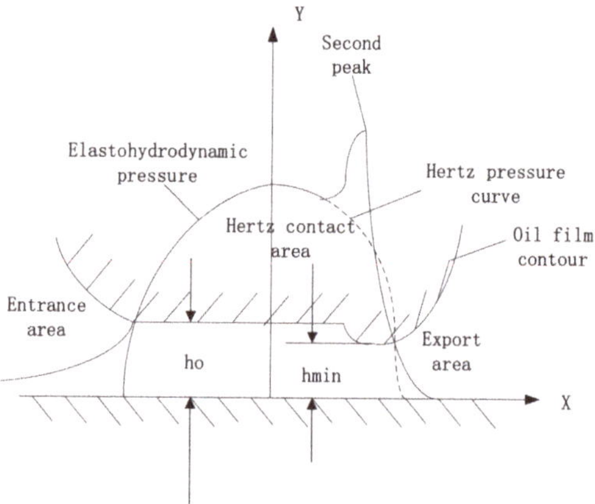

Figure 8. Elasto-hydrodynamic pressure distribution of inter-shaft bearing.

The paper assumed that the lubricating oil is at a constant temperature and that there is no end leakage, and the contact points of rollers and the rings do not slip. The Dowson–Higginson line-contact film-thickness formula is used to calculate the oil-film thickness between the roller and the IR and OR [39] as follows,

$$h_{\min} = \frac{2.56\alpha^{0.54}(\eta_0 U)^{0.7} R^{0.43} L^{0.13}}{E'^{0.03} W^{0.13}} \tag{17}$$

where α is the viscosity coefficient of lubricating oil; η_0 is the dynamic viscosity of lubricating oil; U is the average linear velocity of roller; R is the equivalent radius of curvature; L is the line contact length of roller; W is the radial load; E' is the composite modulus of elasticity, $E' = \frac{E}{1-\nu^2}$.

The inter-shaft bearing is characterized by the same rotation direction of the IR and OR. Assume the rotation speed of the IR is n_i, the radius is R_i, the rotation speed of the OR is n_o, and the roller radius is r. By defining the constant $\gamma = \frac{d}{D_m}\cos\alpha$, the average linear velocity U between the IR and OR of the roller can be obtained according to Equation (5):

$$U = \frac{\pi}{120}D_m[n_i(1-\gamma) + n_0(1+\gamma)] \tag{18}$$

Since the inter-shaft bearing pressure angle $\alpha = 0$, then

$$\gamma = \frac{d}{D_m}\cos\alpha = \frac{r}{R_i + r} \tag{19}$$

The equivalent radius is:

$$R = (\frac{1}{r} \pm \frac{1}{R_i + r \mp r})^{-1} \tag{20}$$

where "$-$" is the contact equivalent radius of the IR and the roller; "$+$" is the contact equivalent radius of the OR and the roller.

Substitute Equation (19) into Equation (20), then:

$$R = r(1 \mp \gamma) \tag{21}$$

By substituting each parameter into Equation (17), the minimum oil-film thickness h_i and h_0 of the rollers and the IR and OR can be obtained.

$$\begin{cases} h_i = 0.21\alpha^{0.54}(\eta_0 D_m)^{0.7}[n_i(1-\gamma) + n_0(1+\gamma)]^{0.7} \\ \quad *r^{0.43}(1-\gamma)^{0.43}L^{0.13}F'^{-0.03}W^{-0.13} \\ h_o = 0.21\alpha^{0.54}(\eta_0 D_m)^{0.7}[n_i(1-\gamma) + n_0(1+\gamma)]^{0.7} \\ \quad *r^{0.43}(1+\gamma)^{0.43}L^{0.13}E'^{-0.03}W^{-0.13} \end{cases} \tag{22}$$

The total thickness of the oil film is:

$$h = h_i + h_0 = C_i W^{-0.13} + C_0 W^{-0.13} \tag{23}$$

where C_i is the coefficient in front of $W^{-0.13}$ in h_i; C_0 is the coefficient in front of $W^{-0.13}$ in h_0.

The oil-film stiffness of the rollers and IR and OR of the inter-shaft bearing can be obtained by Equation (23):

$$K_H = \frac{dW}{dh} = (0.13(C_i + C_0)W^{-1.13})^{-1} \tag{24}$$

According to the calculation formula of series stiffness, the total stiffness of the rolling bearing considering the influence of EHL is:

$$K' = \frac{1}{1/K + 1/K_H} = \frac{KK_H}{K + K_H} \tag{25}$$

3. Experimental Validation of the Numerical Model

3.1. Experimental System

For verifying the validity of the established model, this paper designs and builds an inter-shaft bearing fault simulation test rig. The specific structure of the test rig is shown in Figure 9. The fault simulation experiment system consists of a motor drive, support system, rotor system, and data collection system.

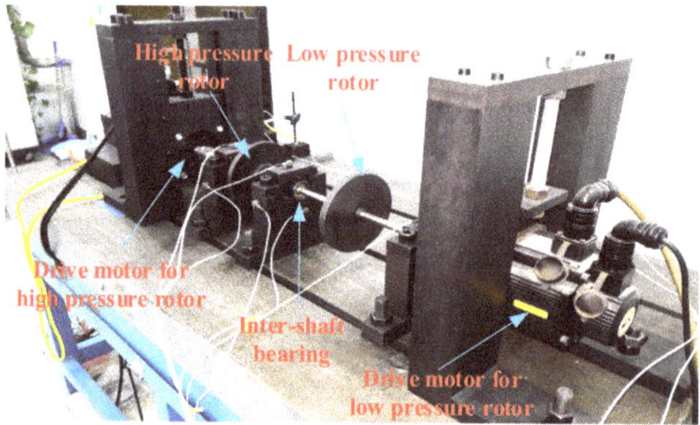

Figure 9. Fault simulation system of inter-shaft bearing.

A NU202EM model from NSK Company (Shenyang, China) was used as an experimental bearing in this paper. Rectangular defects are artificially implanted on the rings and rolling surface of the inter-shaft bearing by the wire-cutting method. The width and depth of the defects on the surface of the IR and OR are 0.5 mm. The width and depth of the defects on the surface of the rollers are 0.1 mm, and the same defects run through the entire cylindrical roller longitudinally. A defective inter-shaft bearing is shown in Figure 10.

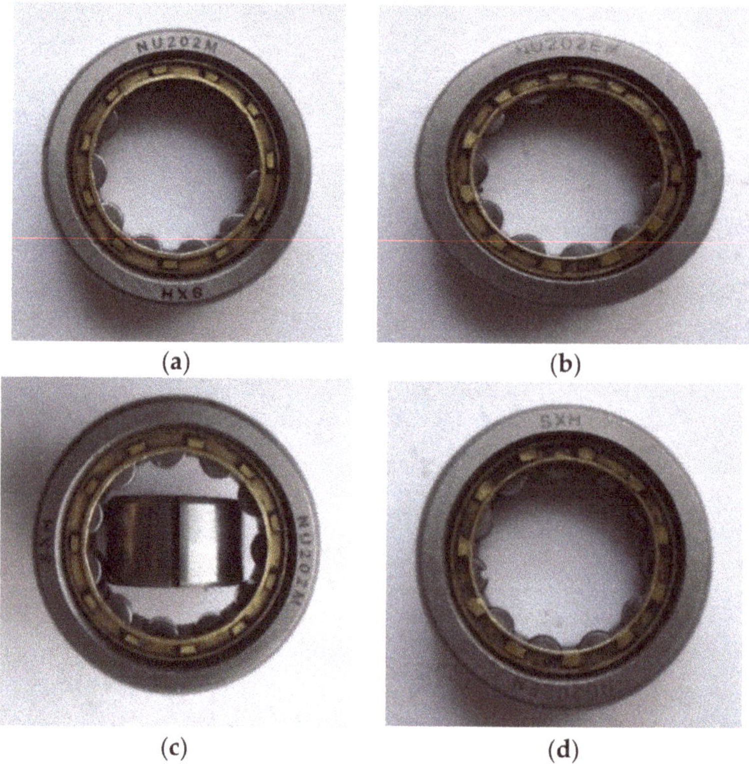

Figure 10. Inter-shaft bearing with local defects. (**a**) Normal statet; (**b**) outer-ring fault; (**c**) inner-ring fault; (**d**) roller fault.

3.2. Bearing Parameters

During the experiment, the IR and OR rotate in the same direction and opposite direction, and the dynamic models of IR fault, OR fault and rollers fault are analyzed respectively. The inter-shaft bearing used is the NU202EM roller bearing. The specific parameters are shown in Table 1.

Table 1. NU202EM bearing parameters.

Parameters	Value
Inner diameter (mm)	15
Outer diameter (mm)	35
Pitch diameter (mm)	25
Roller diameter (mm)	5
Number of rollers	11
Contact angle (°)	0
Radial clearance (μm)	12
Mass (kg)	0.07
Damping coefficient (Ns/m)	300
Bearing width (mm)	11

3.3. Verification and Analysis of Dynamic Model of Inter-Shaft Bearing Fault

Verification of the Simulation Results of the Dynamic Model

(1)   Normal State of Inter-Shaft Bearing

In the experiment, the speed of the OR was set at 1200 r/min and the IR speed was 300 r/min. The radial load was 1000 N and was applied to the inner surface of the IR. The fault characteristic frequency (FCF) of the bearing was calculated according to the empirical formula [10], as shown in Table 2.

Table 2. The FCF of normal state bearing.

Fault Form	Working State	$1 \times f_{zc}$	$2 \times f_{zc}$
Normal state	Counter-rotation	110 Hz	220 Hz

When the inter-shaft bearing is in the normal state, the IR and OR rotate in reverse. Because the vibration signal of the inter-shaft bearing is nonlinear and nonstationary, the traditional Fourier transform can not accurately extract the impact characteristics of faults. Therefore, this paper uses Hilbert envelope analysis to research the vibration signal in the horizontal direction of the inter-shaft bearing [40]. Figure 11 is a time-domain signal and envelope spectrum of the numerical simulation signal of the dynamic model when the inter-shaft bearing has no faults. As can be observed in the time-domain signal, the bearing has no significant impact characteristics. It can be seen from the envelope spectrum that there is only the variable stiffness vibration frequency f_{zc} and its double frequency $2f_{zc}$, and the f_{zc} in the spectrum is 110.2 Hz, which is close to the theoretical calculation value of 110 Hz. Figure 12 shows the time-domain signal and envelope spectrum of the experimental signal of the normal bearing. The characteristic frequency in the envelope spectrum is 114.7 Hz, which is different from the numerical simulation results. Due to the complex structure of the experimental system, the vibration amplitude of the inter-shaft bearing is small under normal conditions, and its signals are easily drowned out by the vibration signals caused by other system components. The bearing is far away from the measuring point, and the energy of the fault vibration signal will be attenuated, which will cause the vibration signal to be submerged by these noises.

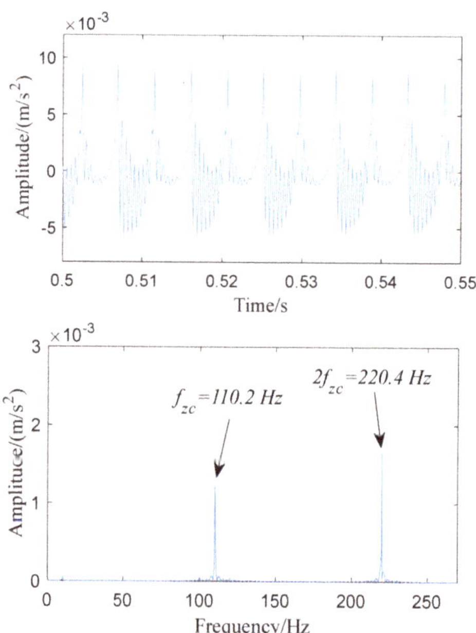

Figure 11. Time–domain signal and envelope spectrum of the normal state (numerical simulation).

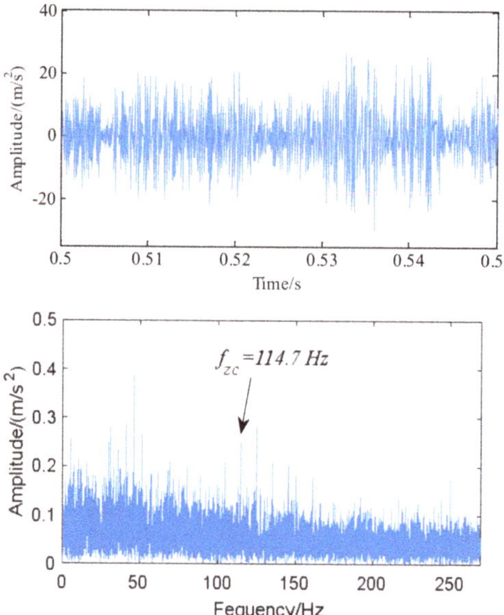

Figure 12. Time–domain signal and envelope spectrum of the normal state (experimental).

(2) Simulation of the Inter-Shaft Bearing with OR Fault

Set the OR rotating speed of the inter-shaft bearing at 300 r/min and the IR speed at 1500 r/min. According to the empirical formula of FCF, the FCF of the OR fault is obtained as shown in Table 3.

Table 3. The FCF of the OR fault.

Fault Form	Working State	$1 \times f_0$	$2 \times f_0$
Outer-ring fault	Counter-rotation	132 Hz	264 Hz

Figure 13 shows the time-domain signal and envelope spectrum of bearing vibration when the OR of the inter-shaft bearing has local defects and the IR and OR rotate in the opposite direction. It can be seen from the time-domain signal that the bearing has obvious impact characteristics. The FCF of OR fault f_0 and its double frequency can be clearly extracted from the envelope spectrum. The rotation frequency of OR F_i an also be extracted. There are also modulation frequencies such as $f_0 \pm F_0$, $f_0 \pm 2F_0$, and $2f_0 \pm F_0$. f_0 is 132.1 Hz, which is only 0.1 Hz different from the theoretical calculation value of 132 Hz. Therefore, it can be proved that the dynamic model established in this paper is accurate and effective for the simulation of reverse rotation.

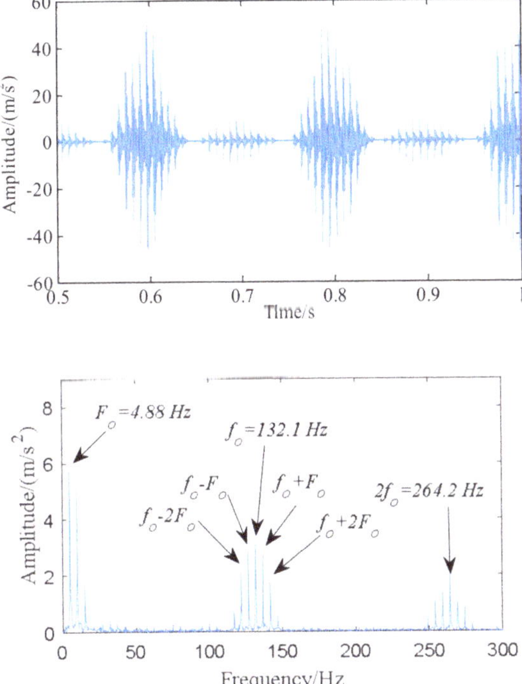

Figure 13. Time–domain signal and envelope spectrum of OR fault (numerical simulation).

Figure 14 shows the time-domain signal and envelope spectrum of the experimental signal. There are also obvious phenomena of shock and amplitude modulation in the collected time-domain signals. The FCF of the OR fault can be clearly extracted as 130.9 Hz in the envelope spectrum, which is 1.2 Hz different from the simulation calculation result but within the allowable error range. This is due to the possibility of rolling element slippage during actual bearing operation. The experimental results also prove that the envelope spectrum contains the FCF of the OR fault f_0 and its multiples component. There are also modulation frequencies such as $f_0 \pm F_0$, $f_0 \pm 2F_0$, and $2f_0 \pm F_0$.

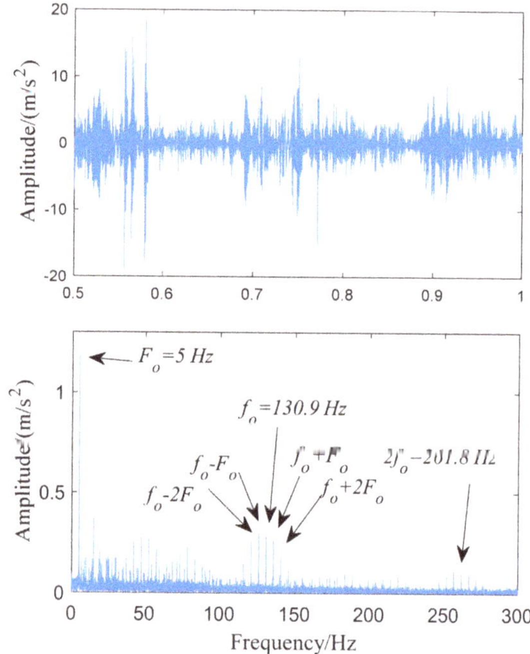

Figure 14. Time–domain signal and envelope spectrum of the OR fault (experimental).

(3)   Simulation of the Inter-Shaft Bearing with the IR Fault

The IR rotation speed of the inter-shaft bearing is set at 300 r/min, and the OR rotation speed is 1200 r/min. According to the theoretical calculation formula, the FCF of the IR fault was calculated as shown in Table 4.

Table 4. The FCF of the IR fault.

Fault Form	Working State	$1 \times f_i$	$2 \times f_i$
Inner-ring fault	Counter-rotation	165 Hz	330 Hz

Figure 15 shows the numerical simulation results of the dynamics model when the IR and OR of the inter-shaft bearing rotate in reverse. The inter-shaft bearing has obvious impact on the time signal. The FCF of the IR fault f_i and its modulation sideband component can be observed in the envelope spectrum. The FCF is calculated as 165.1 Hz by the numerical simulation, which is basically the same as the theoretical value in Table 4. The rotation frequency of IR F_i and $2f_i$ can also be extracted. In the envelope spectrum, it can also be observed that all of the fault frequencies have sideband frequencies with F_i and its multiples as the interval.

Figure 16 is the experimental result of the IR fault, when the IR and OR rotate in reverse. The shock phenomenon of the signal can be clearly seen from the acquired time-domain signal, which is similar to the numerical simulation results. The FCF of the IR fault f_i extracted from experimental signal is 164.8 Hz. Although there is an error between this value and the numerical simulation calculation value of the bearing fault, the error is only 0.3 Hz. The law of signal modulation frequency is consistent with the numerical simulation results. It also proves that the dynamic model of the inter-shaft bearing fault established in this paper is effective and accurate in simulating IR faults.

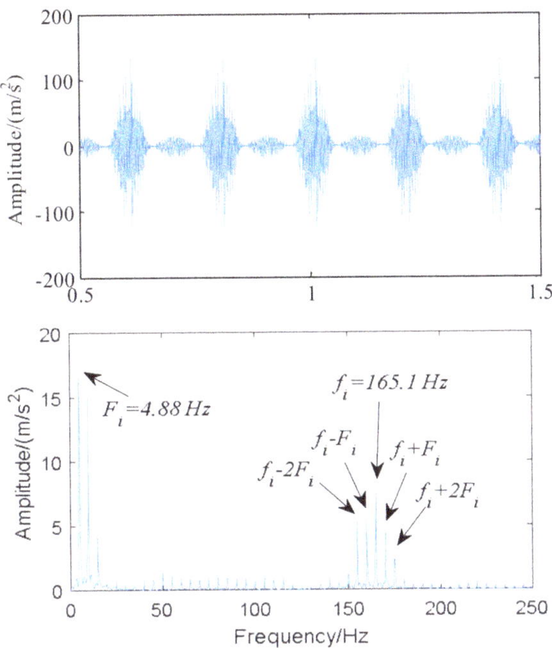

Figure 15. Time–domain signal and envelope spectrum of IR fault (numerical simulation).

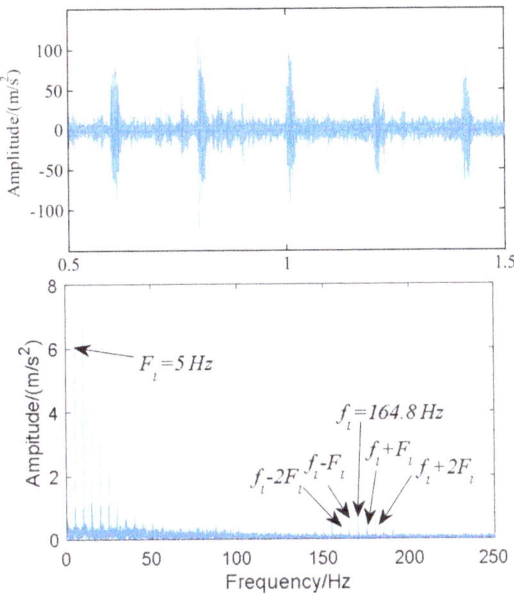

Figure 16. Time–domain signal and envelope spectrum of IR fault (experimental).

(4)   Simulation of the Inter-Shaft Bearing with Roller Fault

The OR rotation speed of the inter-shaft bearing is set at 300 r/min, and the IR rotation speed is 1500 r/min. The FCF of the roller fault are listed in Table 5, when the IR and OR rotate in the opposit direction.

Table 5. The FCF of the roller fault.

Fault Form	Working State	$1 \times f_b$	$2 \times f_b$	Cage Frequency
Roller fault	Counter-rotation	144 Hz	288 Hz	7 Hz

The roller not only rotates around its own axis but also revolves around the bearing center with the cage. When the roller generates a local defect, each rotation of the roller produces two impacts on the IR and OR, and the cage rotation frequency has an amplitude modulation effect on the impact signal. The numerical simulation results of the inter-shaft bearing dynamics model with roller faults are shown in Figure 17. From the envelope spectrum, some data that can be extracted include the FCF of the roller fault FCF f_b and its double frequency, the rotation frequency F_c and its double frequency of the cage, and the sideband frequency with f_b as the center frequency and F_c and its double frequency as the interval. The envelope spectrum shows that the f_b and F_c in the numerical simulation signal of the dynamic model are 144 and 7.02 Hz, respectively. These numerical results is completely consistent with the theoretical calculation results. Figure 18 shows the experimental signal of the roller fault. The obvious shock signal can be seen from the time domain diagram, but the signal components are more complicated. The fault frequency can be extracted from the envelope spectrum, but other frequency components are not obvious.

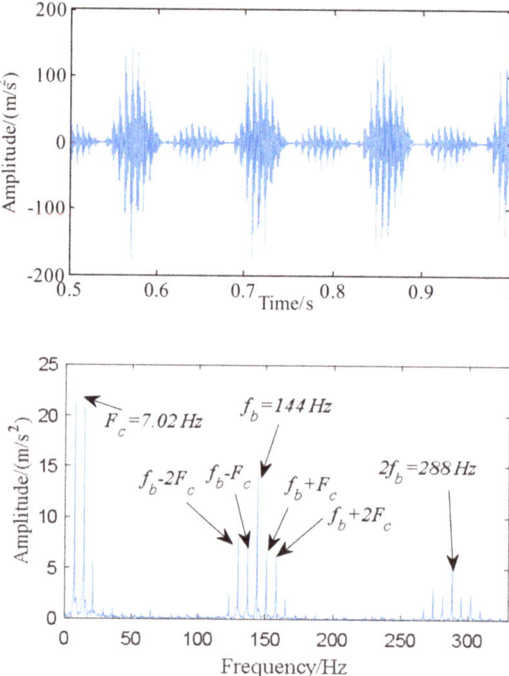

Figure 17. Time–domain signal and envelope spectrum of roller fault (numerical simulation).

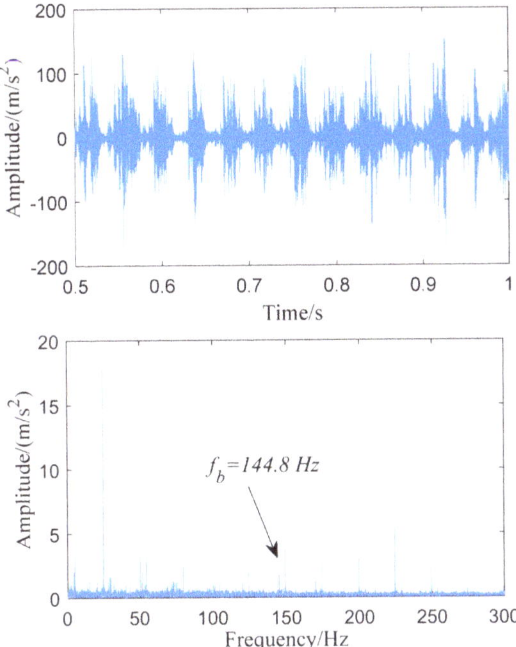

Figure 18. Time–domain signal and envelope spectrum of roller fault (experimental).

Compared with the above three cases of the bearing, the experimental signal envelope spectrum frequency component of the roller fault is different from the simulation signal. The fault signal is seriously interfered with by the noise signal, which makes it difficult to extract fault information. This is also one reason why the research results of such faults are published less frequently.

4. Dynamic Response Analysis of the Inter-Shaft Bearing with Local Defects

4.1. Characteristics of Micro-Local Defects in Inter-Shaft Bearings

Based on the model established above, the time–frequency characteristics of micro faults are analyzed. In the actual working process, it is often necessary to find and locate faults in time when the width of bearing defects is very small. Therefore, based on the established model, this paper studies the time–frequency characteristics of bearing micro faults. When the double rotors rotate in reverse, the vibration response of the OR fault is more obvious than that of the IR fault and roller fault. Therefore, this paper simulates the micro fault in the OR. Set the OR speed at 1000 r/min, the IR speed at 600 r/min, and the simulation results are shown in Figures 19–21.

Figures 19–21 show the simulation results of the time–frequency characteristics when the OR defect size is 0.1, 0.2 and 0.3 mm. When the defect size is 0.1 mm, the time domain signal can also see the impact characteristics, and its vibration amplitude is 2 m/s^2 at most. When the defect size is 0.5 mm in Figure 13, the vibration amplitude has exceeded 40 m/s^2. When the defect sizes are 0.1, 0.2 and 0.3 mm, the time-domain amplitudes of them can be compared. As can be seen from the figure, the vibration amplitude of the bearing rises gradually with the increase in the defect size. Fault is easier to diagnose. When the defect size is 0.1 mm, the FCF of f_0 is very obvious. The sideband frequencies have not appeared. Furthermore, there are a large number of interference frequencies. The above analysis shows that the effect and reliability of the fault feature extraction method based on time–frequency analysis are low for the early micro-fault diagnosis of the inter-shaft bearing.

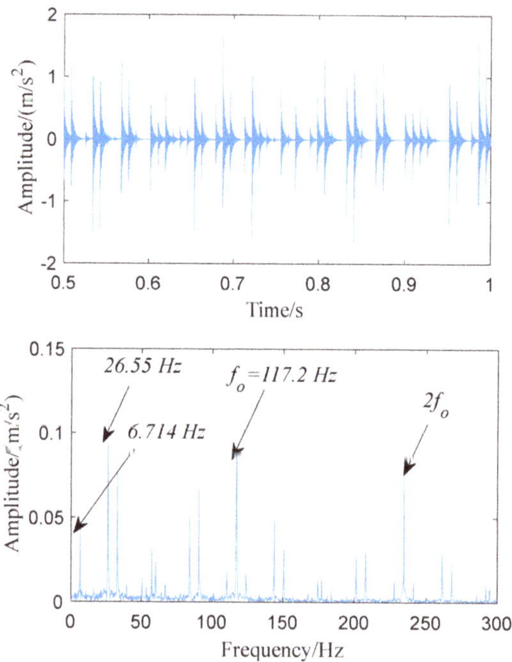

Figure 19. Time–domain waveform and spectrum of the OR fault signal (0.1 mm defect).

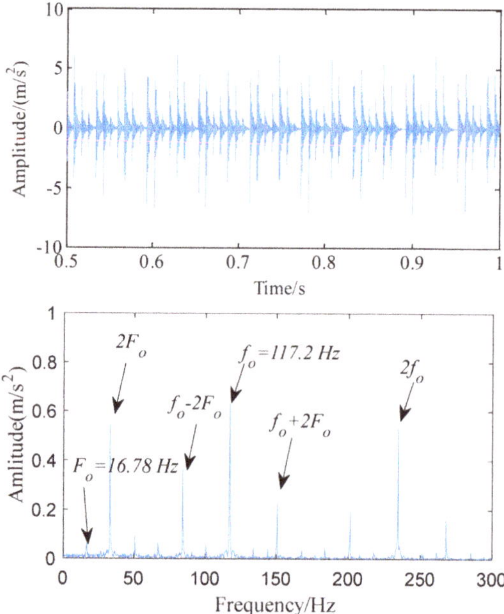

Figure 20. Time–domain waveform and spectrum of the OR fault signal (0.2 mm defect).

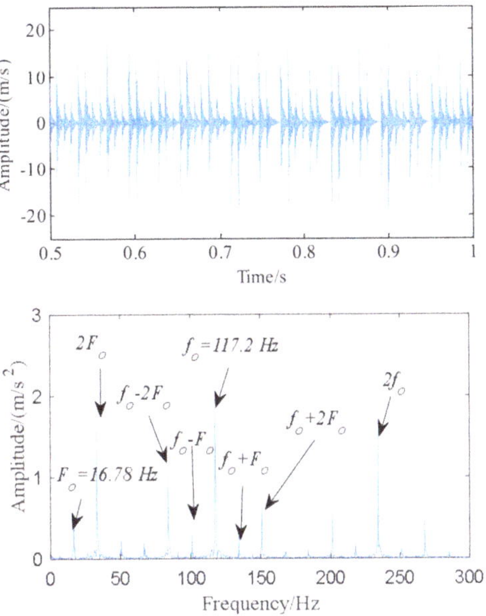

Figure 21. Time–domain waveform and spectrum of the OR fault signal (0.3 mm defect).

4.2. Simulation Analysis of Fault Characteristic Parameters (CP)

Based on a dynamics model with a local defect, the key affecting factors of the fault CP, such as defect width, external load and working speed, are studied. The variation laws of the fault CP are obtained to provide a certain theoretical basis for bearing fault diagnosis and status inspection.

For research on bearing fault diagnosis, the CP, which are susceptible to bearing fault changes, are usually selected to reflect the characteristics of fault signals. CP include both dimensioned CP and dimensionless CP. According to the experience, the dimensioned CP that are sensitive to faults mainly include: maximum value (MV) X_{max}, absolute mean value (AMV) X_a, effective value (EV) X_{rms} and amplitude of square root (AST) X_r. Dimensionless CP include kurtosis factor (KF) K_v, impulse factor (IF) I_f, peak factor (PF) C_f and shape factor (SF) S_f. Set the discrete signal sequence as A, then the calculation formula of fault CP are as follows [41].

$$X_{max} = max(x(k)), k = 1 \dots N \tag{26}$$

$$X_a = \frac{1}{N}\sum_{k=1}^{N}|x(k)| \tag{27}$$

$$X_{rms} = \sqrt{\frac{1}{N}\sum_{k=1}^{N}x^2(k)} \tag{28}$$

$$X_r = \left(\frac{1}{N}\sum_{k=1}^{N}\sqrt{|x(k)|}\right)^2 \tag{29}$$

$$K_v = \frac{\beta}{\sigma^4} \tag{30}$$

$$I_f = \frac{X_{max}}{|\overline{X}|} \tag{31}$$

$$C_f = \frac{X_{max}}{X_{rms}} \tag{32}$$

$$S_f = \frac{X_{rms}}{|\overline{X}|} \tag{33}$$

where β is signal kurtosis, $\beta = \frac{1}{N} \sum_{K=1}^{N} (x(k) - \mu)^4$; σ is the signal standard deviation; μ is the signal mean value.

4.2.1. Effect of Defect Widths on CP

Based on the fault dynamic model of inter-shaft bearing with local defects on the OR, this paper studies the variation law of fault CP with defect widths. Set the radial load at 200 N to remain unchanged, and set the defect widths to change within 0~2 mm. The increment is 0.1 mm. The IR and OR of the inter-shaft bearing rotate in reverse, and the working speed of the IR is 6000 r/min, while that of the OR is 10,000 r/min.

Figure 22 shows the variation of the MV, EV, AST and AMV with the size of defect width. With the increase of defect width, the EV, AST and AMV are increasing gradually. But the increase is relatively smooth and approximately linear, the MV of the signal tends to increase gradually. With the increase of defect width, the fault of the bearing is aggravated, leading to the continuous increase of the vibration amplitude of the bearing. That is why numerical fluctuations in the increasing process and the dimensioned CP vibration parameters tend to increase.

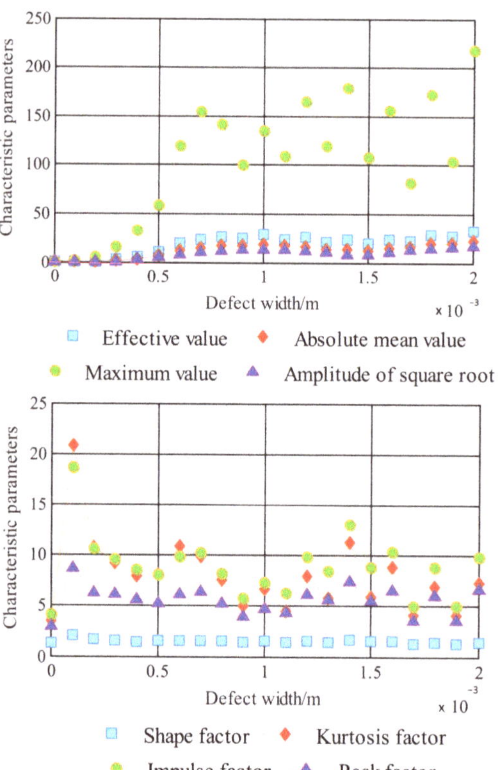

Figure 22. Influence of defect width on dimensional CP.

Figure 23 shows the law of KF, IF, PF and SF of the fault signal with the changing of defect width. With the increase of defect width, KF, IF and PF increased first and then decreased. The SF is not changing significantly with the changing of defect width. It is known from the calculation formula of dimensionless CP that the above-mentioned variation law is caused by different relative growth rates of dimensionless CP on the dimensional CP of the numerator and denominator, with the increase of defect width in different periods of bearing faults.

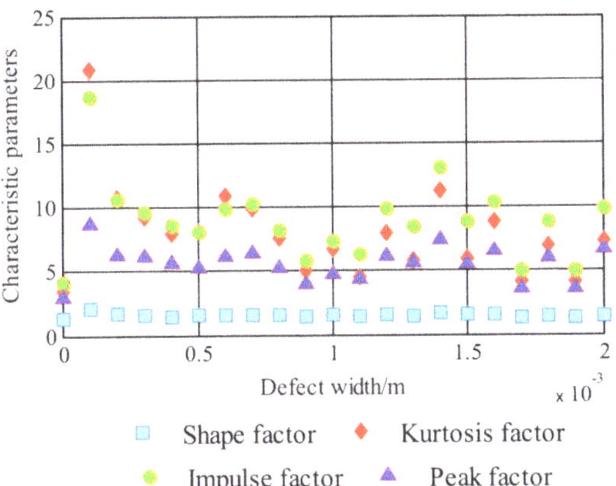

Figure 23. Influence of defect width on dimensionless CP.

4.2.2. Effect of External Loads on CP

The external load variation range is 0–1000 N; the external load increment is 50 N, and the defect width is 2 mm. The IR and OR rotate in the opposite direction. The IR working speed is 6000 r/min, and the OR working speed is 10,000 r/min.

Figure 24 shows the variation laws of the dimensioned CP with the changing of radial loads. With the increases of radial load, EV, AST and AMV tends to increase. This suggests that the statistical parameters of vibration signal can represent the fault characteristics under a radial load. The increase in radial load leads to the decrease of bearing clearance, which makes the roller more easily make contact with the fault. The contact force between the roller and the raceway increases, and the energy of the vibration signal increases accordingly. When the radial load increases, MV firstly increases, then decreases and finally tends to be stable. This indicates that the increase in radial load causes the energy of the bearing vibration to increase; however, the peak of the vibration signal tends to be stable.

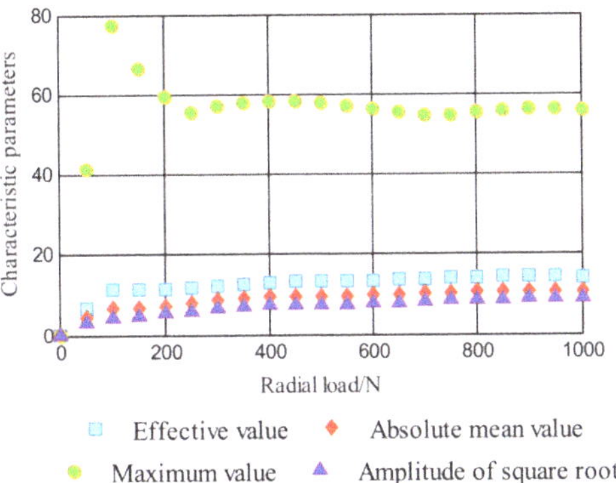

Figure 24. Influence of the radial load on dimensional CP.

Figure 25 shows the variation laws of dimensionless CP with the changing of radial loads. It can be seen that KF and IF is more sensitive to changes in the radial loads. The change trend of the above two parameters is to increase first and then decrease. PF and SF have no obvious change with the radial load, which is not suitable as a CP for fault diagnosis. With the increase in radial loads, the MV increases because the impact energy increases as the roller passes through the defect. However, with the continuous increase in the AMV of the signal, the ratio of them firstly increases, then decreases or tends to be stable.

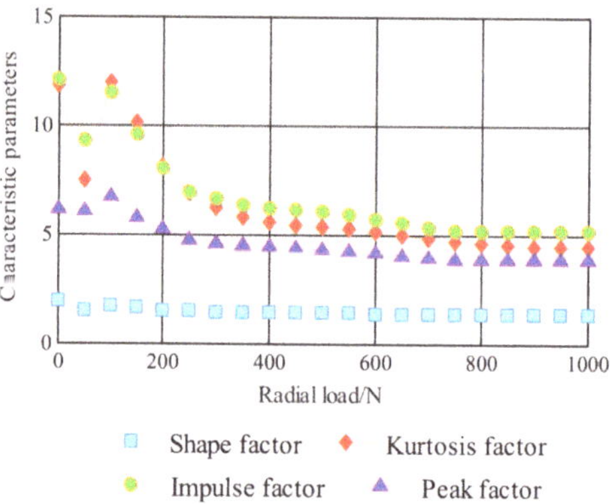

Figure 25. Influence of the radial load on dimensionless CP.

4.2.3. Effect of Rotating Speeds on CP

Set the radial load at 200 N to remain unchanged. The defect width was 2 mm. The IR and OR rotated in reverse. The IR speed was 500 r/min, and the OR speed varied from 500 to 6000 r/min. The increment of speed was 500 r/min.

Figure 26 shows the variation laws of EV, AST, AMV and MV with the changing of speed. Furthermore, all the above dimensional CP tend to increase with the increase of speed. This indicates that dimensional CP can represent the fault characteristics.

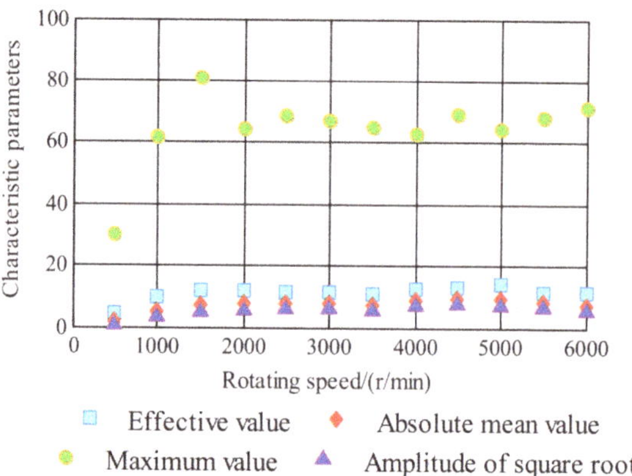

Figure 26. Influence of working speed on dimensional CP.

Figure 27 shows the variation laws of KF, IF, PF and SF with a change in rotating speed. With the increase in rotating speed, SF does not change, but the KF, IF and PF firstly decrease and then increase. This is due to the difference in the relative growth rate of MV and AMV with the increase in rotating speeds.

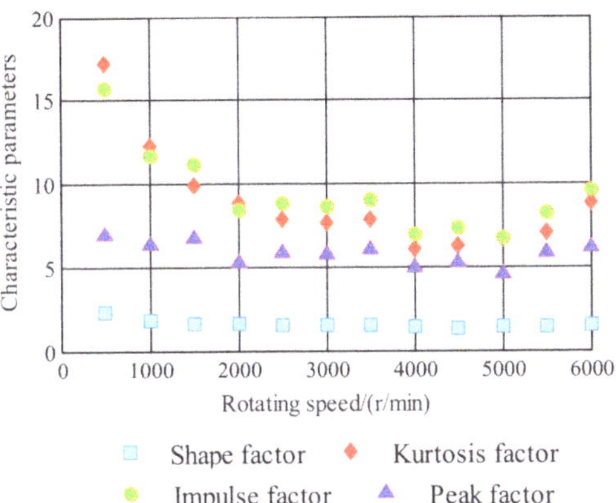

Figure 27. Influence of working speed on dimensionless CP.

5. Conclusions

In order to clarify the fault mechanism of inter-shaft bearings, a 4-DOF local defect inter-shaft bearing of a dynamic model is established in this paper, which considers EHL and TVDE. A piecewise function is used in this model to describe the TVDE of the local defect on the surface of the inter-shaft bearing. At the same time, the influence of the oil film is considered to improve the accuracy of the fault dynamic simulation. In this paper, the frequency distribution of the fault evolution process is simulated and analyzed by the established model. The relationship between defect width, external load and rotating speed and fault CP is studied. The main conclusions of this paper are summarized as follows.

(1) The fault dynamic model established in this paper can simulate the impact characteristics and distribution law of the inter-shaft bearing fault signals accurately, and the fault frequency calculation error is less than 1%.

(2) With the increase of defect size, the vibration amplitude of the inter-shaft bearing also increases. When the width of the defect is less than 0.1 mm, the FCF and a large amount of interference frequently occurs in the envelope spectrum. This can interfere with the early micro-fault diagnosis of the inter-shaft bearing and reduce the effect of the fault diagnosis method based on time–frequency analysis.

(3) The magnitude of the signal peak can reflect the impact force caused by the bearing fault. Normally, the MV increases with the increase in fault impact force since the change of MV is very sensitive to the fault in the early stage of fault; it is also very effective in monitoring the pitting corrosion fault on the bearing surface. AMV, AST and EV can reflect the magnitudes of signal energy. With the development of faults, the overall trend is relatively stable.

(4) The KF and IF indicate whether there is impact component in signals, which is very sensitive to early bearing faults. The PF is sensitive to the change in rotating speed and defect width. The SF is the ratio of the EV to AMV. It has a certain indication effect on the fault, but it is not sensitive to fault change. Therefore, SF is not suitable as the CP for fault diagnosis.

In summary, an effective dynamic model of the inter-shaft bearing is established in this paper to accurately simulate the dynamic characteristics of the inter-shaft bearing with local defects. This study provides useful insights for the use of dynamic models to inspect and monitor the health of inter-shaft bearings of aeroengines.

Author Contributions: Conceptualization, F.Z.; Data curation, Z.W.; Methodology, J.T.; Software, X.A.; Validation, Y.C.; Writing—review & editing, C.W. All authors have read and agreed to the published version of the manuscript.

Funding: This research was funded by National Natural Science Foundation of China (Grant No.12172231), Natural Science Foundation of Liaoning Province of China (Grant No. 2020-BS-174), Liaoning province Department of Education fund (Grant No. JYT2020019) and Research Start-up Funding of Shenyang Aerospace University (Grant No.19YB38). The authors would like to thank them.

Institutional Review Board Statement: Not applicable.

Informed Consent Statement: Not applicable.

Data Availability Statement: Not applicable.

References

1. Lin, S.; Sun, J.; Zhao, C.; Peng, Y. Research on dynamic characteristics of the RBBH system based on dynamics model and vibration data fusion. *Sensors* **2022**, *22*, 3806. [CrossRef] [PubMed]
2. Gupta, P.K. Dynamics of rolling-element bearings-part I: Cylindrical roller bearing analysis. *J. Tribol.* **1979**, *101*, 293–304. [CrossRef]
3. Gupta, P.K. Dynamics of rolling element bearings part II: Cylindrical roller bearing results. *J. Lubr. Technol.* **1979**, *101*, 305–311. [CrossRef]
4. Gupta, P.K. Dynamics of rolling-element bearings part III: Ball bearing analysis. *J. Lubr. Technol.* **1979**, *101*, 312–318. [CrossRef]
5. Gupta, P.K. *Advanced Dynamics of Rolling Elements*, 1st ed.; Springer: New York, NY, USA, 1984.
6. Gupta, P.K.; Dill, J.F.; Bandow, H.E. Dynamics of rolling element bearings experimental validation of the DREB and RAPIDREB computer programs. *J. Tribol.* **1985**, *107*, 132–137. [CrossRef]
7. Gupta, P.K. On the geometrical imperfections in cylindrical roller bearings. *J. Tribol.* **1988**, *110*, 13–18. [CrossRef]
8. Gupta, P.K. On the dynamics of a tapered roller bearing. *J. Tribol.* **1989**, *111*, 278–287. [CrossRef]
9. Gupta, P.K. Analytical modeling of rolling bearings. *Encycl. Tribol.* **2013**, 72–80. [CrossRef]
10. Mcfadden, P.D.; Smith, J.D. Model for the vibration produced by a single point defect in a rolling element bearing. *J. Sound Vib.* **1984**, *96*, 69–82. [CrossRef]
11. Su, Y.T.; Lin, S.J. On initial fault detection of a tapered roller bearing: Frequency domain analysis. *J. Sound Vib.* **1992**, *155*, 75–84. [CrossRef]
12. Su, Y.T.; Lin, M.H.; Lee, M.S. The effects of surface irregularities on roller bearing vibrations. *J. Sound Vib.* **1993**, *165*, 455–466. [CrossRef]
13. Hu, Q.; Deng, S.; Teng, H. A 5-DOF model for aeroengine spindle dual-rotor system analysis. *Chin. J. Aeronaut.* **2011**, *24*, 224–234. [CrossRef]
14. Patel, V.N.; Tandon, N.; Pandey, R.K. Vibration studies of dynamically loaded deep groove ball bearings in presence of local defects on races. *Procedia Eng.* **2013**, *64*, 1582–1591. [CrossRef]
15. Xi, S.T.; Cao, H.R.; Chen, X.F.; Niu, L.K. Dynamic modeling of machine tool spindle bearing system and model based diagnosis of bearing fault caused by collision. *Procedia CIRP* **2018**, *77*, 614–617. [CrossRef]
16. Ma, S.; Li, W.; Yan, K.; Li, Y.; Zhu, Y.; Hong, J. A study on the dynamic contact feature of four-contact-point ball bearing. *Mech. Syst. Signal Process.* **2022**, *174*, 109111. [CrossRef]
17. Patel, U.A.; Upadhyay, S.H. Nonlinear dynamic response of cylindrical roller bearing-rotor system with 9 degree of freedom model having a combined localized defect at inner-outer races of bearing. *Tribol. Trans.* **2016**, *60*, 284–299. [CrossRef]
18. Liu, J.; Xue, L.; Xu, Z.; Wu, H.; Guang, P. Vibration characteristics of a high-speed flexible angular contact ball bearing with the manufacturing error. *Mech. Mach. Theory* **2021**, *162*, 104335. [CrossRef]
19. Cao, H.; Niu, L.; He, Z. Method for vibration response simulation and sensor placement optimization of a machine tool spindle system with a bearing defect. *Sensors* **2012**, *12*, 8732–8754. [CrossRef]
20. Patil, M.S.; Mathew, J.; Rajendrakumar, P.K.; Desai, S. A Theoretical model to predict the effect of localized defect on vibrations associated with ball bearing. *Int. J. Mech. Sci.* **2010**, *52*, 1193–1201. [CrossRef]
21. Kulkarni, P.G.; Sahasrabudhe, A.D. A dynamic model of ball bearing for simulating localized defects on outer race using cubic hermite spline. *J. Mech. Sci. Technol.* **2014**, *28*, 3433–3442. [CrossRef]
22. Cui, L.L.; Zhang, Y.; Zhang, F.B.; Lee, S.C. Vibration response mechanism of faulty outer race rolling element bearings for quantitative analysis. *J. Sound Vib.* **2016**, *364*, 67–76. [CrossRef]
23. Khanam, S.; Tandon, N.; Dutt, J.K. Multi-event excitation force model for inner race defect in a rolling element bearing. *J. Tribol.* **2016**, *138*, 011106. [CrossRef]
24. Wang, F.T.; Jing, M.Q.; Yi, J.; Dong, G.H. Dynamic modelling for vibration analysis of a cylindrical roller bearing due to localized defects on raceways. *Proc. Inst. Mech. Eng. Part K J. Multi-Body Dyn.* **2015**, 22939–22964. [CrossRef]
25. Sassi, S.; Badri, B.; Thomas, M. A numerical model to predict damaged bearing vibrations. *J. Vib. Control.* **2007**, *13*, 1603–1628. [CrossRef]

26. Wang, C.; Wu, J.; Cheng, J.; Zhang, S.; Zhang, K. Elasto-hydrodynamic lubrication of the journal bearing system with a relief groove in the scroll compressor: Simulation and experiment. *Tribol. Int.* **2022**, *165*, 107252. [CrossRef]
27. Yan, C.F.; Yuan, H.; Wang, X.; Wu, L.X. Dynamics modeling on local defect of deep groove ball bearing under point contact elasto-hydrodynamic lubrication condition. *J. Vib. Shock* **2016**, *35*, 61–70.
28. Zhang, Y.Y.; Wang, X.L.; Yan, X.L. Dynamic behaviors of the elasto-hydrodynamic lubricated contact for rolling bearings. *J. Tribol.* **2012**, *135*, 021501. [CrossRef]
29. Choudhury, A.; Tandon, N. Vibration response of rolling element bearings in a rotor bearing system to a local defect under radial load. *J. Tribol.* **2006**, *128*, 252–261. [CrossRef]
30. Jiang, Y.C.; Huang, W.T.; Luo, J.N.; Wang, W.J. An improved dynamic model of defective bearings considering the three-dimensional geometric relationship between the rolling element and defect area. *Mech. Syst. Signal Process.* **2019**, *129*, 694–716. [CrossRef]
31. Bourdon, A.; Chesné, S.; André, H.; Rémond, D. Reconstruction of angular speed variations in the angular domain to diagnose and quantify taper roller bearing outer race fault. *Mech. Syst. Signal Process.* **2019**, *120*, 1–15. [CrossRef]
32. Li, Y.M.; Cao, H.R.; Tang, K. A general dynamic model coupled with EFEM and DBM of rolling bearing-rotor system. *Mech. Syst. Signal Process.* **2019**, *134*, 106322. [CrossRef]
33. Niu, L.K.; Cao, H.R.; Hou, H.P.; Wu, B.; Lan, Y.; Xiong, X.Y. Experimental observations and dynamic modeling of vibration characteristics of a cylindrical roller bearing with roller defects. *Mech. Syst. Signal Process.* **2020**, *138*, 106553. [CrossRef]
34. Cao, H.R.; Shi, F.; Li, Y.M.; Li, B.J.; Chen, X.F. Vibration and stability analysis of rotor-bearing-pedestal system due to clearance fit. *Mech. Syst. Signal Process.* **2019**, *133*, 106275. [CrossRef]
35. Lei, Y.; Liu, Z.; Wang, D.; Yang, X.; Liu, H.; Lin, J. A probability distribution model of tooth pits for evaluating time-varying mesh stiffness of pitting gears. *Mech. Syst. Signal Process.* **2018**, *106*, 355–366. [CrossRef]
36. Harris, T.A.; Kotzalas, M.N. *Advanced Concepts of Bearing Technology: Rolling Bearing Analysis*, 5th ed.; Crc Press: Boca Raton, FL, USA, 2007.
37. Shi, S.-B.; Shao, W.; Wei, X.-K.; Yang, X.-S.; Wang, B.-Z. A New unconditionally stable FDTD method based on the newmark-beta algorithm. *IEEE Trans. Microw. Theory Tech.* **2016**, *64*, 4082–4090. [CrossRef]
38. Liu, J.; Ni, H.; Xu, Z.; Pan, G. A simulation analysis for lubricating characteristics of an oil-jet lubricated ball bearing. *Simul. Model. Pract. Theory* **2021**, *113*, 102371. [CrossRef]
39. Lubrecht, A.A.; Venner, C.H.; Colin, F. Film thickness calculation in elasto-hydrodynamic lubricated line and elliptical contacts: The Dowson, Higginson, Hamrock contribution. *Proc. Inst. Mech. Eng. Part J J. Eng. Tribol.* **2009**, *223*, 511–515. [CrossRef]
40. Tian, J.; Ai, Y.-T.; Fei, C.-W.; Zhang, F.-L.; Choy, Y.-S. Dynamic modeling and simulation analysis on outer race fault of inter-shaft bearing. *J. Propuls. Technol.* **2019**, *40*, 660–666.
41. Wang, F.T.; Sun, J.; Yan, D.W.; Zhang, S.H.; Cui, L.M.; Xu, Y. A feature extraction method for fault classification of rolling bearing based on PCA. *J. Phys. Conf. Ser.* **2015**, *628*, 012079. [CrossRef]

MDPI

St. Alban-Anlage 66

4052 Basel

Switzerland

Tel. +41 61 683 77 34

Fax +41 61 302 89 18

www.mdpi.com

Coatings Editorial Office

E-mail: coatings@mdpi.com

www.mdpi.com/journal/coatings